# FREE Test Taking Tips DVD Offer

To help us better serve you, we have developed a Test Taking Tips DVD that we would like to give you for FREE. **This DVD covers world-class test taking tips that you can use to be even more successful when you are taking your test.**

All that we ask is that you email us your feedback about your study guide. Please let us know what you thought about it – whether that is good, bad or indifferent.

To get your **FREE Test Taking Tips DVD**, email freedvd@studyguideteam.com with "FREE DVD" in the subject line and the following information in the body of the email:

> a. The title of your study guide.
>
> b. Your product rating on a scale of 1-5, with 5 being the highest rating.
>
> c. Your feedback about the study guide. What did you think of it?
>
> d. Your full name and shipping address to send your free DVD.

If you have any questions or concerns, please don't hesitate to contact us at freedvd@studyguideteam.com.

Thanks again!

# SAT Math Prep 2021 and 2022

## Study Book with 3 Practice Tests
## [2nd Edition]

Joshua Rueda

Written and edited by TPB Publishing.

TPB Publishing is not associated with or endorsed by any official testing organization. TPB Publishing is a publisher of unofficial educational products. All test and organization names are trademarks of their respective owners. Content in this book is included for utilitarian purposes only and does not constitute an endorsement by TPB Publishing of any particular point of view.

Interested in buying more than 10 copies of our product? Contact us about bulk discounts:
bulkorders@studyguideteam.com

ISBN 13: 9781637755723
ISBN 10: 1637755724

# Table of Contents

# Quick Overview

As you draw closer to taking your exam, effective preparation becomes more and more important. Thankfully, you have this study guide to help you get ready. Use this guide to help keep your studying on track and refer to it often.

This study guide contains several key sections that will help you be successful on your exam. The guide contains tips for what you should do the night before and the day of the test. Also included are test-taking tips. Knowing the right information is not always enough. Many well-prepared test takers struggle with exams. These tips will help equip you to accurately read, assess, and answer test questions.

A large part of the guide is devoted to showing you what content to expect on the exam and to helping you better understand that content. In this guide are practice test questions so that you can see how well you have grasped the content. Then, answer explanations are provided so that you can understand why you missed certain questions.

Don't try to cram the night before you take your exam. This is not a wise strategy for a few reasons. First, your retention of the information will be low. Your time would be better used by reviewing information you already know rather than trying to learn a lot of new information. Second, you will likely become stressed as you try to gain a large amount of knowledge in a short amount of time. Third, you will be depriving yourself of sleep. So be sure to go to bed at a reasonable time the night before. Being well-rested helps you focus and remain calm.

Be sure to eat a substantial breakfast the morning of the exam. If you are taking the exam in the afternoon, be sure to have a good lunch as well. Being hungry is distracting and can make it difficult to focus. You have hopefully spent lots of time preparing for the exam. Don't let an empty stomach get in the way of success!

When travelling to the testing center, leave earlier than needed. That way, you have a buffer in case you experience any delays. This will help you remain calm and will keep you from missing your appointment time at the testing center.

Be sure to pace yourself during the exam. Don't try to rush through the exam. There is no need to risk performing poorly on the exam just so you can leave the testing center early. Allow yourself to use all of the allotted time if needed.

Remain positive while taking the exam even if you feel like you are performing poorly. Thinking about the content you should have mastered will not help you perform better on the exam.

Once the exam is complete, take some time to relax. Even if you feel that you need to take the exam again, you will be well served by some down time before you begin studying again. It's often easier to convince yourself to study if you know that it will come with a reward!

# Test-Taking Strategies

## 1. Predicting the Answer

When you feel confident in your preparation for a multiple-choice test, try predicting the answer before reading the answer choices. This is especially useful on questions that test objective factual knowledge. By predicting the answer before reading the available choices, you eliminate the possibility that you will be distracted or led astray by an incorrect answer choice. You will feel more confident in your selection if you read the question, predict the answer, and then find your prediction among the answer choices. After using this strategy, be sure to still read all of the answer choices carefully and completely. If you feel unprepared, you should not attempt to predict the answers. This would be a waste of time and an opportunity for your mind to wander in the wrong direction.

## 2. Reading the Whole Question

Too often, test takers scan a multiple-choice question, recognize a few familiar words, and immediately jump to the answer choices. Test authors are aware of this common impatience, and they will sometimes prey upon it. For instance, a test author might subtly turn the question into a negative, or he or she might redirect the focus of the question right at the end. The only way to avoid falling into these traps is to read the entirety of the question carefully before reading the answer choices.

## 3. Looking for Wrong Answers

Long and complicated multiple-choice questions can be intimidating. One way to simplify a difficult multiple-choice question is to eliminate all of the answer choices that are clearly wrong. In most sets of answers, there will be at least one selection that can be dismissed right away. If the test is administered on paper, the test taker could draw a line through it to indicate that it may be ignored; otherwise, the test taker will have to perform this operation mentally or on scratch paper. In either case, once the obviously incorrect answers have been eliminated, the remaining choices may be considered. Sometimes identifying the clearly wrong answers will give the test taker some information about the correct answer. For instance, if one of the remaining answer choices is a direct opposite of one of the eliminated answer choices, it may well be the correct answer. The opposite of obviously wrong is obviously right! Of course, this is not always the case. Some answers are obviously incorrect simply because they are irrelevant to the question being asked. Still, identifying and eliminating some incorrect answer choices is a good way to simplify a multiple-choice question.

## 4. Don't Overanalyze

Anxious test takers often overanalyze questions. When you are nervous, your brain will often run wild, causing you to make associations and discover clues that don't actually exist. If you feel that this may be a problem for you, do whatever you can to slow down during the test. Try taking a deep breath or counting to ten. As you read and consider the question, restrict yourself to the particular words used by the author. Avoid thought tangents about what the author *really* meant, or what he or she was *trying* to say. The only things that matter on a multiple-choice test are the words that are actually in the question. You must avoid reading too much into a multiple-choice question, or supposing that the writer meant something other than what he or she wrote.

## 5. No Need for Panic

It is wise to learn as many strategies as possible before taking a multiple-choice test, but it is likely that you will come across a few questions for which you simply don't know the answer. In this situation, avoid panicking. Because most multiple-choice tests include dozens of questions, the relative value of a single wrong answer is small. As much as possible, you should compartmentalize each question on a multiple-choice test. In other words, you should not allow your feelings about one question to affect your success on the others. When you find a question that you either don't understand or don't know how to answer, just take a deep breath and do your best. Read the entire question slowly and carefully. Try rephrasing the question a couple of different ways. Then, read all of the answer choices carefully. After eliminating obviously wrong answers, make a selection and move on to the next question.

## 6. Confusing Answer Choices

When working on a difficult multiple-choice question, there may be a tendency to focus on the answer choices that are the easiest to understand. Many people, whether consciously or not, gravitate to the answer choices that require the least concentration, knowledge, and memory. This is a mistake. When you come across an answer choice that is confusing, you should give it extra attention. A question might be confusing because you do not know the subject matter to which it refers. If this is the case, don't eliminate the answer before you have affirmatively settled on another. When you come across an answer choice of this type, set it aside as you look at the remaining choices. If you can confidently assert that one of the other choices is correct, you can leave the confusing answer aside. Otherwise, you will need to take a moment to try to better understand the confusing answer choice. Rephrasing is one way to tease out the sense of a confusing answer choice.

## 7. Your First Instinct

Many people struggle with multiple-choice tests because they overthink the questions. If you have studied sufficiently for the test, you should be prepared to trust your first instinct once you have carefully and completely read the question and all of the answer choices. There is a great deal of research suggesting that the mind can come to the correct conclusion very quickly once it has obtained all of the relevant information. At times, it may seem to you as if your intuition is working faster even than your reasoning mind. This may in fact be true. The knowledge you obtain while studying may be retrieved from your subconscious before you have a chance to work out the associations that support it. Verify your instinct by working out the reasons that it should be trusted.

## 8. Key Words

Many test takers struggle with multiple-choice questions because they have poor reading comprehension skills. Quickly reading and understanding a multiple-choice question requires a mixture of skill and experience. To help with this, try jotting down a few key words and phrases on a piece of scrap paper. Doing this concentrates the process of reading and forces the mind to weigh the relative importance of the question's parts. In selecting words and phrases to write down, the test taker thinks about the question more deeply and carefully. This is especially true for multiple-choice questions that are preceded by a long prompt.

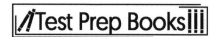

## 9. Subtle Negatives

One of the oldest tricks in the multiple-choice test writer's book is to subtly reverse the meaning of a question with a word like *not* or *except*. If you are not paying attention to each word in the question, you can easily be led astray by this trick. For instance, a common question format is, "Which of the following is…?" Obviously, if the question instead is, "Which of the following is not…?," then the answer will be quite different. Even worse, the test makers are aware of the potential for this mistake and will include one answer choice that would be correct if the question were not negated or reversed. A test taker who misses the reversal will find what he or she believes to be a correct answer and will be so confident that he or she will fail to reread the question and discover the original error. The only way to avoid this is to practice a wide variety of multiple-choice questions and to pay close attention to each and every word.

## 10. Reading Every Answer Choice

It may seem obvious, but you should always read every one of the answer choices! Too many test takers fall into the habit of scanning the question and assuming that they understand the question because they recognize a few key words. From there, they pick the first answer choice that answers the question they believe they have read. Test takers who read all of the answer choices might discover that one of the latter answer choices is actually *more* correct. Moreover, reading all of the answer choices can remind you of facts related to the question that can help you arrive at the correct answer. Sometimes, a misstatement or incorrect detail in one of the latter answer choices will trigger your memory of the subject and will enable you to find the right answer. Failing to read all of the answer choices is like not reading all of the items on a restaurant menu: you might miss out on the perfect choice.

## 11. Spot the Hedges

One of the keys to success on multiple-choice tests is paying close attention to every word. This is never truer than with words like almost, most, some, and sometimes. These words are called "hedges" because they indicate that a statement is not totally true or not true in every place and time. An absolute statement will contain no hedges, but in many subjects, the answers are not always straightforward or absolute. There are always exceptions to the rules in these subjects. For this reason, you should favor those multiple-choice questions that contain hedging language. The presence of qualifying words indicates that the author is taking special care with his or her words, which is certainly important when composing the right answer. After all, there are many ways to be wrong, but there is only one way to be right!  For this reason, it is wise to avoid answers that are absolute when taking a multiple-choice test. An absolute answer is one that says things are either all one way or all another. They often include words like *every*, *always*, *best*, and *never*. If you are taking a multiple-choice test in a subject that doesn't lend itself to absolute answers, be on your guard if you see any of these words.

## 12. Long Answers

In many subject areas, the answers are not simple. As already mentioned, the right answer often requires hedges. Another common feature of the answers to a complex or subjective question are qualifying clauses, which are groups of words that subtly modify the meaning of the sentence. If the question or answer choice describes a rule to which there are exceptions or the subject matter is complicated, ambiguous, or confusing, the correct answer will require many words in order to be expressed clearly and accurately. In essence, you should not be deterred by answer choices that seem excessively long. Oftentimes, the author of the text will not be able to write the correct answer without

offering some qualifications and modifications. Your job is to read the answer choices thoroughly and completely and to select the one that most accurately and precisely answers the question.

## 13. Restating to Understand

Sometimes, a question on a multiple-choice test is difficult not because of what it asks but because of how it is written. If this is the case, restate the question or answer choice in different words. This process serves a couple of important purposes. First, it forces you to concentrate on the core of the question. In order to rephrase the question accurately, you have to understand it well. Rephrasing the question will concentrate your mind on the key words and ideas. Second, it will present the information to your mind in a fresh way. This process may trigger your memory and render some useful scrap of information picked up while studying.

## 14. True Statements

Sometimes an answer choice will be true in itself, but it does not answer the question. This is one of the main reasons why it is essential to read the question carefully and completely before proceeding to the answer choices. Too often, test takers skip ahead to the answer choices and look for true statements. Having found one of these, they are content to select it without reference to the question above. Obviously, this provides an easy way for test makers to play tricks. The savvy test taker will always read the entire question before turning to the answer choices. Then, having settled on a correct answer choice, he or she will refer to the original question and ensure that the selected answer is relevant. The mistake of choosing a correct-but-irrelevant answer choice is especially common on questions related to specific pieces of objective knowledge. A prepared test taker will have a wealth of factual knowledge at his or her disposal, and should not be careless in its application.

## 15. No Patterns

One of the more dangerous ideas that circulates about multiple-choice tests is that the correct answers tend to fall into patterns. These erroneous ideas range from a belief that B and C are the most common right answers, to the idea that an unprepared test-taker should answer "A-B-A-C-A-D-A-B-A." It cannot be emphasized enough that pattern-seeking of this type is exactly the WRONG way to approach a multiple-choice test. To begin with, it is highly unlikely that the test maker will plot the correct answers according to some predetermined pattern. The questions are scrambled and delivered in a random order. Furthermore, even if the test maker was following a pattern in the assignation of correct answers, there is no reason why the test taker would know which pattern he or she was using. Any attempt to discern a pattern in the answer choices is a waste of time and a distraction from the real work of taking the test. A test taker would be much better served by extra preparation before the test than by reliance on a pattern in the answers.

# FREE DVD OFFER

Don't forget that doing well on your exam includes both understanding the test content and understanding how to use what you know to do well on the test. We offer a completely FREE Test Taking Tips DVD that covers world class test taking tips that you can use to be even more successful when you are taking your test.

All that we ask is that you email us your feedback about your study guide. To get your **FREE Test Taking Tips DVD**, email freedvd@studyguideteam.com with "FREE DVD" in the subject line and the following information in the body of the email:

- The title of your study guide.
- Your product rating on a scale of 1-5, with 5 being the highest rating.
- Your feedback about the study guide. What did you think of it?
- Your full name and shipping address to send your free DVD.

# Introduction to the SAT

## Function of the Test

The SAT is a standardized test taken by high school students across the United States and given internationally for college placement. It is designed to measure problem solving ability, communication, and understanding complex relationships. The SAT also serves as a qualifying measure to identify students for college scholarships, depending on the college being applied to. All colleges in the U.S. accept the SAT, and, in addition to admissions and scholarships, use SAT scores for course placement as well as academic counseling.

Most of the high school students who take the SAT are seniors. In 2016, the number of students who took the SAT was just under 1.7 million. It's important to note that since many updates have been implemented during the 2016 year, the data points cannot be compared to those in previous years. In 2014, 42.6 percent of students met the College Board's "college and career readiness" benchmark, and in 2015, 41.9 percent met this benchmark.

## Test Administration

The SAT is offered on seven days throughout the year at schools throughout the United States. Internationally, the SAT is offered on five days throughout the year. There are thousands of testing centers worldwide. Test-takers can view the test centers in their area when they register for the test, or they can view testing locations at the College Board website, a not-for-profit that owns and publishes the SAT.

The SAT registration fee is $46, and the SAT with Essay registration fee is $60, although both of these have fee waivers available. Also note that students outside the U.S. may have to pay an extra processing fee. Additional fees include registering by phone, changing fee, late registration fee, or a waitlist fee. Test-takers may register four score reports for free up to nine days after the test. Any additional score reports cost $12, although fee waivers are available for this as well.

## Test Format

The SAT gauges a student's proficiency in three areas: Reading, Mathematics, and Writing and Language. The reading portion of the SAT measures comprehension, requiring candidates to read multi-paragraph fiction and non-fiction segments including informational visuals, such as charts, tables and graphs, and answer questions based on this content. Fluency in problem solving, conceptual understanding of equations, and real-world applications are characteristics of the math test. The writing and language portion requires students to evaluate and edit writing and graphics to obtain an answer that correctly conveys the information given in the passage.

The SAT contains 154 multiple-choice questions, with each section comprising over 40 questions. A different length of time is given for each section, for a total of three hours, plus fifty minutes for the essay (optional).

| Section | Time (In Minutes) | Number of Questions |
|---|---|---|
| Reading | 65 | 52 |
| Writing and Language | 35 | 44 |
| Mathematics | 80 | 58 |
| Essay (optional) | 50 (optional) | 1 (optional) |
| **Total** | **180** | **154** |

## Scoring

Scores for the new SAT are based on a scale from 400 to 1600. Scores range from 200 to 800 for Evidence-Based Reading and Writing, and 200 to 800 for Math. The optional essay is scored from 2 to 8. The SAT also no longer penalizes for incorrect answers. Therefore, a student's raw score is the number of correctly answered questions.

On the College Board website, there are indicators to determine what the benchmark scores are. The scores are divided up into green, yellow, or red. Green meets or exceeds the benchmark, and shows a 480 to 800 in Evidence-Based Reading and Writing, and a 530 to 800 in Math.

## Recent/Future Developments

The SAT taken before March 2016 is different than the one administered currently. Currently, the essay is optional, and the time limit for the Reading, Writing, and Math sections has increased per section. The content features also vary, with the new test focusing on skills that research has identified as most important for college readiness and the meaning of words in extended context rather than emphasis on vocabulary. The score range has changed from 600–2400 to 400–1600. There has also been added subscore reporting, which provides insight to students and parents about the scores.

# SAT Mathematics

The Math section of the SAT focuses on a variety of math concepts and practices including real world applications of math operations and relations. The test includes a total of 58 questions to be answered in 80 minutes. The test contains multiple choice type questions and grid in questions where the test taker will be asked to fill in the answer rather than selecting from a number of options. The test provides instructions on how to complete the grid in questions. Test takers must be careful to only mark one circle in each column of the grid. Answers written only in the boxes above the circles will not be counted. If a fraction is an answer to a grid in question, they need not be simplified, but they must be converted to improper fractions. The Math Test also includes both a calculator section and a no calculator section. Test takers are not allowed to use calculators on the no calculator section; this is to assess math fluency, use of number operations, and number sense.

The Math section will ask the test taker to demonstrate their abilities in problem solving and modeling that reflect the types of situations that students will encounter in future college courses, careers, and everyday life. The Math section is broken down into three major content sections. They are the heart of algebra, problem solving and data analysis, and passport to advanced math. The test also incorporates additional topics in math that will help prepare students for their college and career.

The heart of algebra section encompasses the study of algebra and the major concepts needed to solve and create linear equations and functions. Students must also use analytical and problem solving skills to solve questions concerning linear inequalities and systems of equations. Questions in this section are presented in a variety of ways including graphical and algebraic representations of problems that require different strategies and processes to complete.

The problem solving and data analysis section presents questions involving units and quantities that must be manipulated and converted using ratios, rates, and proportional relationships. Test takers will also need to interpret graphs, charts, and other representations of data to identify key features of a data set such as measures of center, patterns, spread, and deviations.

The passport to advanced math section covers topics that will help ready students for more advanced math topics. One of the major topics in this section is knowledge of expressions and how to work with them. Also included in this section are building functions and interpreting complex equations.

The additional topics in math have two main areas: geometric and trigonometric concepts. These include, but are not limited to, volume formulas, trigonometric ratios, and equations and theorems related to circles.

Scores for the Math section of the SAT range from 200 to 800. The three main areas of the Math section, heart of algebra, problem solving and data analysis, and passport to advanced math, are also given subscores which range from 1 to 15.

# Heart of Algebra

## Solving Linear Equations

A function is called *linear* if it can take the form of the equation $f(x) = ax + b$, or $y = ax + b$, for any two numbers $a$ and $b$. A linear equation forms a straight line when graphed on the coordinate plane. An example of a linear function is shown below on the graph.

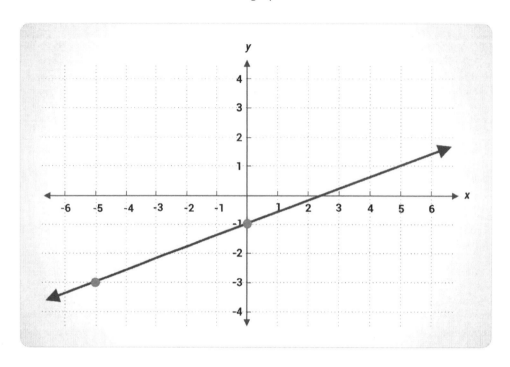

This is a graph of the following function: $y = \frac{2}{5}x - 1$. A table of values that satisfies this function is shown below.

| x | y |
|---|---|
| -5 | -3 |
| 0 | -1 |
| 5 | 1 |
| 10 | 3 |

These points can be found on the graph using the form (x, y).

When graphing a linear function, note that the ratio of the change of the *y* coordinate to the change in the *x* coordinate is constant between any two points on the resulting line, no matter which two points are chosen. In other words, in a pair of points on a line, $(x_1, y_1)$ and $(x_2, y_2)$, with $x_1 \neq x_2$ so that the two points are distinct, then the ratio $\frac{y_2 - y_1}{x_2 - x_1}$ will be the same, regardless of which particular pair of points are chosen. This ratio, $\frac{y_2 - y_1}{x_2 - x_1}$, is called the *slope* of the line and is frequently denoted with the letter $m$. If slope $m$ is positive, then the line goes upward when moving to the right, while if slope $m$ is

negative, then the line goes downward when moving to the right. If the slope is 0, then the line is called *horizontal*, and the *y* coordinate is constant along the entire line. In lines where the *x* coordinate is constant along the entire line, *y* is not actually a function of *x*. For such lines, the slope is not defined. These lines are called *vertical* lines.

Linear functions may take forms other than $y = ax + b$. The most common forms of linear equations are explained below:

1. Standard Form: $Ax + By = C$, in which the slope is given by $m = \frac{-A}{B}$, and the *y*-intercept is given by $\frac{C}{B}$.

2. Slope-Intercept Form: $y = mx + b$, where the slope is *m* and the *y* intercept is *b*.

3. Point-Slope Form: $y - y_1 = m(x - x_1)$, where the slope is *m* and $(x_1, y_1)$ is any point on the chosen line.

4. Two-Point Form: $\frac{y-y_1}{x-x_1} = \frac{y_2-y_1}{x_2-x_1}$, where $(x_1, y_1)$ and $(x_2, y_2)$ are any two distinct points on the chosen line. Note that the slope is given by $m = \frac{y_2-y_1}{x_2-x_1}$.

5. Intercept Form: $\frac{x}{x_1} + \frac{y}{y_1} = 1$, in which $x_1$ is the *x*-intercept and $y_1$ is the *y*-intercept.

These five ways to write linear equations are all useful in different circumstances. Depending on the given information, it may be easier to write one of the forms over another.

If $y = mx$, *y* is directly proportional to *x*. In this case, changing *x* by a factor changes *y* by that same factor. If $y = \frac{m}{x}$, *y* is inversely proportional to *x*. For example, if *x* is increased by a factor of 3, then *y* will be decreased by the same factor, 3.

Sometimes, rather than a situation where there's an equation such as $y = ax + b$ and finding *y* for some value of *x* is requested, the result is given and finding *x* is requested.

The key to solving any equation is to remember that from one true equation, another true equation can be found by adding, subtracting, multiplying, or dividing both sides by the same quantity. In this case, it's necessary to manipulate the equation so that one side only contains *x*. Then the other side will show what *x* is equal to.

For example, in solving $3x - 5 = 2$, adding 5 to each side results in $3x = 7$. Next, dividing both sides by 3 results in $x = \frac{7}{3}$. To ensure the value of x is correct, the number can be substituted into the original equation and solved to see if it makes a true statement. For example, $3(\frac{7}{3}) - 5 = 2$ can be simplified by cancelling out the two 3s. This yields $7 - 5 = 2$, which is a true statement.

Sometimes an equation may have more than one *x* term. For example, consider the following equation:

$$3x + 2 = x - 4$$

Moving all of the *x* terms to one side by subtracting *x* from both sides results in:

$$2x + 2 = -4$$

Next, subtract 2 from both sides so that there is no constant term on the left side. This yields $2x = -6$. Finally, divide both sides by 2, which leaves $x = -3$.

## Solving Linear Inequalities

Solving linear inequalities is very similar to solving equations, except for one rule: when multiplying or dividing an inequality by a negative number, the inequality symbol changes direction. Given the following inequality, solve for x: $-2x + 5 < 13$. The first step in solving this equation is to subtract 5 from both sides. This leaves the inequality: $-2x < 8$. The last step is to divide both sides by -2. By using the rule, the answer to the inequality is $x > -4$.

Since solutions to inequalities include more than one value, number lines are used many times to model the answer. For the previous example, the answer is modelled on the number line below. It shows that any number greater than -4, not including -4, satisfies the inequality.

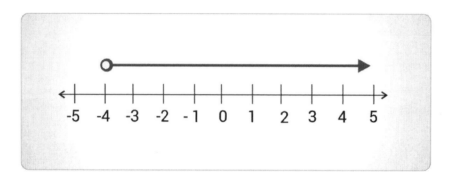

Similar to linear equations, a linear inequality may have a solution set consisting of all real numbers, or can contain no solution. When solved algebraically, a linear inequality in which the variable cancels out and results in a true statement (ex. $7 \geq 2$) has a solution set of all real numbers. A linear inequality in which the variable cancels out and results in a false statement (ex. $7 \leq 2$) has no solution.

## Building Linear Functions

A *function* is a special kind of relation where, for each value of x, there is only a single value of y that satisfies the relation. So, $x^2 = y^2$ is *not* a function because in this case, if x is 1, y can be either 1 or -1: the pair (1, 1) and (1, -1) both satisfy the relation. More generally, for this relation, any pair of the form $(a, \pm a)$ will satisfy it. On the other hand, consider the following relation:

$$y = x^2 + 1$$

This is a function because for each value of x, there is a unique value of y that satisfies the relation. Notice, however, there are multiple values of x that give us the same value of y. This is perfectly acceptable for a function. Therefore, y is a function of x.

To determine if a relation is a function, check to see if every x value has a unique corresponding y value.

A function can be viewed as an object that has x as its input and outputs a unique y-value. It is sometimes convenient to express this using *function notation*, where the function itself is given a name,

often $f$. To emphasize that $f$ takes $x$ as its input, the function is written as $f(x)$. In the above example, the equation could be rewritten as:

$$f(x) = x^2 + 1$$

To write the value that a function yields for some specific value of $x$, that value is put in place of $x$ in the function notation. For example, $f(3)$ means the value that the function outputs when the input value is 3. If $f(x) = x^2 + 1$, then:

$$f(3) = 3^2 + 1 = 10$$

A function can also be viewed as a table of pairs $(x, y)$, which lists the value for $y$ for each possible value of $x$.

The set of all possible values for $x$ in $f(x)$ is called the *domain* of the function, and the set of all possible outputs is called the *range* of the function. Note that usually the domain is assumed to be all real numbers, except those for which the expression for $f(x)$ is not defined, unless the problem specifies otherwise. An example of how a function might not be defined is in the case of $f(x) = \frac{1}{x+1}$, which is not defined when $x = -1$ (which would require dividing by zero). Therefore, in this case the domain would be all real numbers except $x = -1$.

If $y$ is a function of $x$, then $x$ is the *independent variable* and $y$ is the *dependent variable*. This is because in many cases, the problem will start with some value of $x$ and then see how $y$ changes depending on this starting value.

Functions can be built out of the context of a situation. For example, the relationship between the money paid for a gym membership and the months that someone has been a member can be described through a function. If the one-time membership fee is \$40 and the monthly fee is \$30, then the function can be written:

$$f(x) = 30x + 40$$

The $x$-value represents the number of months the person has been part of the gym, while the output is the total money paid for the membership. The table below shows this relationship. It is a representation of the function because the initial cost is \$40 and the cost increases each month by \$30.

| x (months) | y (money paid to gym) |
|---|---|
| 0 | 40 |
| 1 | 70 |
| 2 | 100 |
| 3 | 130 |

Functions can also be built from existing functions. For example, a given function $f(x)$ can be transformed by adding a constant, multiplying by a constant, or changing the input value by a constant. The new function $g(x) = f(x) + k$ represents a vertical shift of the original function. In $f(x) = 3x - 2$, a vertical shift 4 units up would be:

$$g(x) = 3x - 2 + 4 = 3x + 2$$

Multiplying the function times a constant $k$ represents a vertical stretch, based on whether the constant is greater than or less than 1. The function

$$g(x) = kf(x) = 4(3x - 2) = 12x - 8$$

represents a stretch. Changing the input $x$ by a constant forms the function:

$$g(x) = f(x + k) = 3(x + 4) - 2 = 3x + 12 - 2 = 3x + 10$$

and this represents a horizontal shift to the left 4 units. If $(x - 4)$ was plugged into the function, it would represent a vertical shift.

To evaluate functions, plug in the given value everywhere the variable appears in the expression for the function. For example, find $g(-2)$ where:

$$g(x) = 2x^2 - \frac{4}{x}$$

To complete the problem, plug in -2 in the following way:

$$g(-2) = 2(-2)^2 - \frac{4}{-2} = 2 \cdot 4 + 2 = 8 + 2 = 10$$

## Solving Systems of Inequalities

A linear inequality in two variables is a statement expressing an unequal relationship between those two variables. Typically written in slope-intercept form, the variable $y$ can be greater than, less than, greater than or equal to, or less than or equal to a linear expression that includes the variable $x$, such as $y > 3x$ and $y \leq \frac{1}{2}x - 3$. Questions may include instructions to model real-world scenarios, such as the following:

You work part time cutting lawns for $15 each and cleaning houses for $25 each. Your goal is to make more than $90 this week. Write an inequality to represent the possible pairs of lawns and houses needed to reach your goal.

This scenario can be expressed as $15x + 25y > 90$ where $x$ is the number of lawns cut and $y$ is the number of houses cleaned.

The graph consists of a boundary line dividing the coordinate plane and shading on one side of the boundary. Graph the boundary line just as a linear equation would be graphed. If the inequality symbol is > or <, use a dashed line to indicate that the line is not part of the solution set. If the inequality symbol is ≥ or ≤, use a solid line to indicate that the boundary line is included in the solution set. Pick an ordered pair $(x, y)$ on either side of the line to test in the inequality statement. If substituting the values for $x$ and $y$ results in a true statement $[15(3) + 25(2) > 90]$, that ordered pair and all others on that side of the boundary line are part of the solution set. To indicate this, shade that region of the graph. If substituting the ordered pair results in a false statement, the ordered pair and all others on that side are

not part of the solution set. Therefore, the other region of the graph contains the solutions and should be shaded. The following is an example of the graph of:

$$y \leq x + 2$$

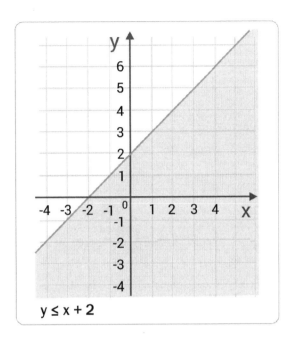

$y \leq x + 2$

Systems of *linear inequalities* are like systems of equations, but the solutions are different. Since inequalities have infinitely many solutions, their systems also have infinitely many solutions. Finding the solutions of inequalities involves graphs. A system of two equations and two inequalities is linear; thus, the lines can be graphed using slope-intercept form. A system of linear inequalities consists of two linear inequalities that make comparisons between two variables. The solution set for a system of inequalities is the region of a graph consisting of ordered pairs that make BOTH inequalities true. To graph the solution set, first graph each linear inequality with appropriate shading. Identify the region of the graph where the shading for the two inequalities overlaps. This region contains the solution set for the system. In the example below, the line with the positive slope is solid, meaning the values on that line are included in the solution. The line with the negative slope is dotted, so the coordinates on that line are not included.

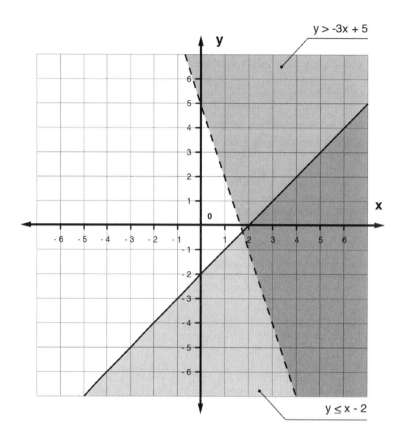

## Solving Systems of Linear Equations

A *system of equations* is a group of equations that have the same variables or unknowns. These equations can be linear, but they are not always so. Finding a solution to a system of equations means finding the values of the variables that satisfy each equation. For a linear system of two equations and two variables, there could be a single solution, no solution, or infinitely many solutions.

A single solution occurs when there is one value for $x$ and y that satisfies the system. This would be shown on the graph where the lines cross at exactly one point. When there is no solution, the lines are parallel and do not ever cross. With infinitely many solutions, the equations may look different, but they are the same line. One equation will be a multiple of the other, and on the graph, they lie on top of each other.

The process of elimination can be used to solve a system of equations. For example, the following equations make up a system:

$$x + 3y = 10 \text{ and } 2x - 5y = 9$$

Immediately adding these equations does not eliminate a variable, but it is possible to change the first equation by multiplying the whole equation by $-2$. This changes the first equation to

$$-2x - 6y = -20$$

The equations can be then added to obtain $-11y = -11$. Solving for $y$ yields $y = 1$. To find the rest of the solution, 1 can be substituted in for $y$ in either original equation to find the value of $x = 7$. The solution to the system is (7, 1) because it makes both equations true, and it is the point in which the lines intersect. If the system is *dependent*—having infinitely many solutions—then both variables will cancel out when the elimination method is used, resulting in an equation that is true for many values of $x$ and $y$. Since the system is dependent, both equations can be simplified to the same equation or line.

A system can also be solved using *substitution*. This involves solving one equation for a variable and then plugging that solved equation into the other equation in the system. This equation can be solved for one variable, which can then be plugged in to either original equation and solved for the other variable. For example, $x - y = -2$ and $3x + 2y = 9$ can be solved using substitution. The first equation can be solved for $x$, where:

$$x = -2 + y$$

Then it can be plugged into the other equation:

$$3(-2 + y) + 2y = 9$$

Solving for $y$ yields:

$$-6 + 3y + 2y = 9$$

That shows that $y = 3$. If $y = 3$, then $x = 1$.

This solution can be checked by plugging in these values for the variables in each equation to see if it makes a true statement.

Finally, a solution to a system of equations can be found graphically. The solution to a linear system is the point or points where the lines cross. The values of x and y represent the coordinates $(x, y)$ where the lines intersect. Using the same system of equation as above, they can be solved for $y$ to put them in slope-intercept form,

$$y = mx + b$$

These equations become $y = x + 2$ and:

$$y = -\frac{3}{2}x + 4.5$$

The slope is the coefficient of $x$, and the y-intercept is the constant value.

This system with the solution is shown below:

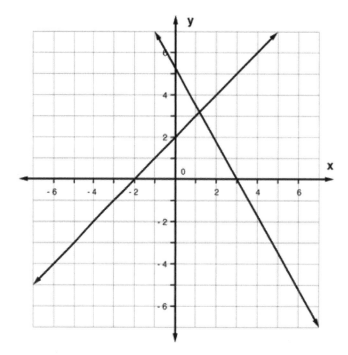

Finding solutions to systems of equations is essentially finding what values of the variables make both equations true. It is finding the input value that yields the same output value in both equations. For functions $g(x)$ and $f(x)$, the equation $g(x) = f(x)$ means the output values are being set equal to each other. Solving for the value of $x$ means finding the $x$-coordinate that gives the same output in both functions. For example, $f(x) = x + 2$ and $g(x) = -3x + 10$ is a system of equations. Setting $f(x) = g(x)$ yields the equation:

$$x + 2 = -3x + 10$$

Solving for $x$, gives the $x$-coordinate $x = 2$ where the two lines cross. This value can also be found by using a table or a graph. On a table, both equations can be given the same inputs, and the outputs can be recorded to find the point(s) where the lines cross. Any method of solving finds the same solution, but some methods are more appropriate for some systems of equations than others.

## Interpreting Variables and Constants in Expressions

Algebraic expressions look similar to equations, but they do not include the equal sign. Algebraic expressions are comprised of numbers, variables, and mathematical operations. Some examples of algebraic expressions are $8x + 7y - 12z$, $3a^2$, and $5x^3 - 4y^4$.

Algebraic expressions and equations can be used to represent real-life situations and model the behavior of different variables. For example, $2x + 5$ could represent the cost to play games at an arcade. In this case, 5 represents the price of admission to the arcade, and 2 represents the cost of each game played. To calculate the total cost, use the number of games played for x, multiply it by 2, and add 5.

In word problems, multiple quantities are often provided with a request to find some kind of relation between them. This often will mean that one variable (the dependent variable whose value needs to be

found) can be written as a function of another variable (the independent variable whose value can be figured from the given information). The usual procedure for solving these problems is to start by giving each quantity in the problem a variable, and then figuring the relationship between these variables.

For example, suppose a car gets 25 miles per gallon. How far will the car travel if it uses 2.4 gallons of fuel? In this case, $y$ would be the distance the car has traveled in miles, and $x$ would be the amount of fuel burned in gallons (2.4). Then the relationship between these variables can be written as an algebraic equation, $y = 25x$. In this case, the equation is $y = 25 \cdot 2.4 = 60$, so the car has traveled 60 miles.

Some word problems require more than just one simple equation to be written and solved. Consider the following situations and the linear equations used to model them.

Suppose Margaret is 2 miles to the east of John at noon. Margaret walks to the east at 3 miles per hour. How far apart will they be at 3 p.m.? To solve this, $x$ would represent the time in hours past noon, and $y$ would represent the distance between Margaret and John. Now, noon corresponds to the equation where $x$ is 0, so the $y$ intercept is going to be 2. It's also known that the slope will be the rate at which the distance is changing, which is 3 miles per hour. This means that the slope will be 3 (be careful at this point: if units were used, other than miles and hours, for $x$ and $y$ variables, a conversion of the given information to the appropriate units would be required first). The simplest way to write an equation given the $y$-intercept, and the slope is the Slope-Intercept form, is $y = mx + b$. Recall that $m$ here is the slope, and b is the $y$ intercept. So, $m = 3$ and $b = 2$. Therefore, the equation will be:

$$y = 3x + 2$$

The word problem asks how far to the east Margaret will be from John at 3 p.m., which means when $x$ is 3. So, substitute $x = 3$ into this equation to obtain:

$$y = 3 \times 3 + 2 = 9 + 2 = 11$$

Therefore, she will be 11 miles to the east of him at 3 p.m.

For another example, suppose that a box with 4 cans in it weighs 6 lbs., while a box with 8 cans in it weighs 12 lbs. Find out how much a single can weighs. To do this, let $x$ denote the number of cans in the box, and $y$ denote the weight of the box with the cans in lbs. This line touches two pairs: $(4, 6)$ and $(8, 12)$. A formula for this relation could be written using the two-point form, with:

$$x_1 = 4, y_1 = 6, x_2 = 8, y_2 = 12$$

This would yield $\frac{y-6}{x-4} = \frac{12-6}{8-4}$, or:

$$\frac{y - 6}{x - 4} = \frac{6}{4} = \frac{3}{2}$$

However, only the slope is needed to solve this problem, since the slope will be the weight of a single can. From the computation, the slope is $\frac{3}{2}$. Therefore, each can weighs $\frac{3}{2}$ lb.

## Understanding Connections Between Algebraic and Graphical Representations

To graph relations and functions, the Cartesian plane is used. This means to think of the plane as being given a grid of squares, with one direction being the $x$-axis and the other direction the $y$-axis. Generally, the independent variable is placed along the horizontal axis, and the dependent variable is placed along the vertical axis. Any point on the plane can be specified by saying how far to go along the x-axis and how far along the y-axis with a pair of numbers $(x, y)$. Specific values for these pairs can be given names such as $C = (-1, 3)$. Negative values mean to move left or down; positive values mean to move right or up. The point where the axes cross one another is called the *origin*. The origin has coordinates $(0, 0)$ and is usually called $O$ when given a specific label.

An illustration of the Cartesian plane, along with graphs of $(2, 1)$ and $(-1, -1)$, are below.

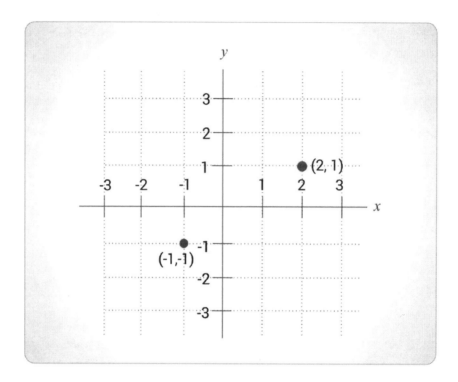

Relations also can be graphed by marking each point whose coordinates satisfy the relation. If the relation is a function, then there is only one value of *y* for any given value of *x*. This leads to the **vertical line test**: if a relation is graphed, then the relation is a function if any possible vertical line drawn anywhere along the graph would only touch the graph of the relation in no more than one place. Conversely, when graphing a function, then any possible vertical line drawn will not touch the graph of the function at any point or will touch the function at just one point. This test is made from the definition of a function, where each *x*-value must be mapped to one and only one y-value.

Equations and inequalities in two variables represent a relationship. Jim owns a car wash and charges $40 per car. The rent for the facility is $350 per month. An equation can be written to relate the number of cars Jim cleans to the money he makes per month. Let $x$ represent the number of cars and $y$ represent the profit Jim makes each month from the car wash. The equation $y = 40x - 350$ can be used to show Jim's profit or loss. Since this equation has two variables, the coordinate plane can be used to show the relationship and predict profit or loss for Jim.

The following graph shows that Jim must wash at least nine cars to pay the rent, where $x = 9$. Anything nine cars and above yield a profit shown in the value on the y-axis.

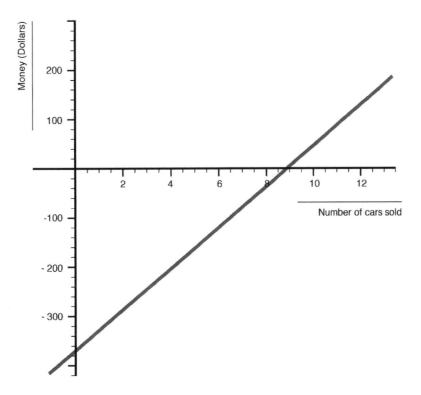

With a single equation in two variables, the solutions are limited only by the situation the equation represents. When two equations or inequalities are used, more constraints are added. For example, in a system of linear equations, there is often—although not always—only one answer. The point of intersection of two lines is the solution. For a system of inequalities, there are infinitely many answers.

# Problem Solving and Data Analysis

## Ratios, Rates, and Proportions

*Ratios* are used to show the relationship between two quantities. The ratio of oranges to apples in the grocery store may be 3 to 2. That means that for every 3 oranges, there are 2 apples. This comparison can be expanded to represent the actual number of oranges and apples. Another example may be the number of boys to girls in a math class. If the ratio of boys to girls is given as 2 to 5, that means there are 2 boys to every 5 girls in the class. Ratios can also be compared if the units in each ratio are the same. The ratio of boys to girls in the math class can be compared to the ratio of boys to girls in a science class by stating which ratio is higher and which is lower.

Rates are used to compare two quantities with different units. *Unit rates* are the simplest form of rate. With unit rates, the denominator in the comparison of two units is one. For example, if someone can type at a rate of 1000 words in 5 minutes, then his or her unit rate for typing is $\frac{1000}{5} = 200$ words in one minute or 200 words per minute. Any rate can be converted into a unit rate by dividing to make the denominator one. 1000 words in 5 minutes has been converted into the unit rate of 200 words per minute.

Ratios and rates can be used together to convert rates into different units. For example, if someone is driving 50 kilometers per hour, that rate can be converted into miles per hour by using a ratio known as the *conversion factor*. Since the given value contains kilometers and the final answer needs to be in miles, the ratio relating miles to kilometers needs to be used. There are 0.62 miles in 1 kilometer. This, written as a ratio and in fraction form, is

$$\frac{0.62\ miles}{1\ km}$$

To convert 50km/hour into miles per hour, the following conversion needs to be set up:

$$\frac{50\ km}{hour} \times \frac{0.62\ miles}{1\ km} = 31\ miles\ per\ hour$$

The ratio between two similar geometric figures is called the *scale factor*. For example, a problem may depict two similar triangles, A and B. The scale factor from the smaller triangle A to the larger triangle B is given as 2 because the length of the corresponding side of the larger triangle, 16, is twice the corresponding side on the smaller triangle, 8. This scale factor can also be used to find the value of a missing side, $x$, in triangle A. Since the scale factor from the smaller triangle (A) to larger one (B) is 2, the larger corresponding side in triangle B (given as 25), can be divided by 2 to find the missing side in A ($x = 12.5$). The scale factor can also be represented in the equation $2A = B$ because two times the lengths of A gives the corresponding lengths of B. This is the idea behind similar triangles.

Much like a scale factor can be written using an equation like $2A = B$, a *relationship* is represented by the equation $Y = kX$. X and Y are proportional because as values of X increase, the values of Y also increase. A relationship that is inversely proportional can be represented by the equation $Y = \frac{k}{x}$, where the value of Y decreases as the value of $x$ increases and vice versa.

Proportional reasoning can be used to solve problems involving ratios, percentages, and averages. Ratios can be used in setting up proportions and solving them to find unknowns. For example, if a student completes an average of 10 pages of math homework in 3 nights, how long would it take the student to complete 22 pages? Both ratios can be written as fractions. The second ratio would contain the unknown.

The following proportion represents this problem, where x is the unknown number of nights:

$$\frac{10\ pages}{3\ nights} = \frac{22\ pages}{x\ nights}$$

Solving this proportion entails cross-multiplying and results in the following equation:

$$10x = 22 \times 3$$

Simplifying and solving for $x$ results in the exact solution: $x = 6.6\ nights$. The result would be rounded up to 7 because the homework would actually be completed on the 7th night.

The following problem uses ratios involving percentages:

If 20% of the class is girls and 30 students are in the class, how many girls are in the class?

To set up this problem, it is helpful to use the common proportion:

$$\frac{\%}{100} = \frac{is}{of}$$

Within the proportion, % is the percentage of girls, 100 is the total percentage of the class, *is* is the number of girls, and *of* is the total number of students in the class. Most percentage problems can be written using this language. To solve this problem, the proportion should be set up as $\frac{20}{100} = \frac{x}{30}$, and then solved for x. Cross-multiplying results in the equation $20 \times 30 = 100x$, which results in the solution $x = 6$. There are 6 girls in the class.

Problems involving volume, length, and other units can also be solved using ratios. A problem may ask for the volume of a cone to be found that has a radius, $r = 7m$ and a height, $h = 16m$. Referring to the formulas provided on the test, the volume of a cone is given as:

$$V = \pi r^2 \frac{h}{3}$$

r is the radius, and h is the height. Plugging $r = 7$ and $h = 16$ into the formula, the following is obtained:

$$V = \pi (7^2) \frac{16}{3}$$

Therefore, volume of the cone is found to be approximately 821m³. Sometimes, answers in different units are sought. If this problem wanted the answer in liters, 821m³ would need to be converted.

Using the equivalence statement 1m³ = 1000L, the following ratio would be used to solve for liters:

$$821\text{m}^3 \times \frac{1000L}{1m^3}$$

Cubic meters in the numerator and denominator cancel each other out, and the answer is converted to 821,000 liters, or $8.21 * 10^5$ L.

Other conversions can also be made between different given and final units. If the temperature in a pool is 30°C, what is the temperature of the pool in degrees Fahrenheit? To convert these units, an equation is used relating Celsius to Fahrenheit. The following equation is used:

$$T_{°F} = 1.8 T_{°C} + 32$$

Plugging in the given temperature and solving the equation for T yields the result:

$$T_{°F} = 1.8(30) + 32 = 86°F$$

Both units in the metric system and U.S. customary system are widely used.

## Percentages

The word *percent* comes from the Latin phrase for "per one hundred." A *percent* is a way of writing out a fraction. It is a fraction with a denominator of 100. Thus, $65\% = \frac{65}{100}$.

To convert a fraction to a percent, the denominator is written as 100. For example, $\frac{3}{5} = \frac{60}{100} = 60\%$.

In converting a percent to a fraction, the percent is written with a denominator of 100, and the result is simplified. For example, $30\% = \frac{30}{100} = \frac{3}{10}$.

The basic percent equation is the following:

$$\frac{is}{of} = \frac{\%}{100}$$

The placement of numbers in the equation depends on what the question asks.

Example 1
Find 40% of 80.

Basically, the problem is asking, "What is 40% of 80?" The 40% is the percent, and 80 is the number to find the percent "of." The equation is:

$$\frac{x}{80} = \frac{40}{100}$$

Solving the equation by cross-multiplication, the problem becomes 100x = 80(40). Solving for x gives the answer: $x = 32$.

Example 2
What percent of 100 is 20?

The 20 fills in the "is" portion, while 100 fills in the "of." The question asks for the percent, so that will be x, the unknown. The following equation is set up:

$$\frac{20}{100} = \frac{x}{100}$$

Cross-multiplying yields the equation 100x = 20(100). Solving for x gives the answer of 20%.

Example 3
30% of what number is 30?

The following equation uses the clues and numbers in the problem:

$$\frac{30}{x} = \frac{30}{100}$$

Cross-multiplying results in the equation 30(100) = 30x. Solving for x gives the answer x = 100.

## Unit Rates and Conversions

When rates are expressed as a quantity of one, they are considered unit rates. To determine a unit rate, the first quantity is divided by the second. Knowing a unit rate makes calculations easier than simply

having a rate. For example, suppose a 3-pound bag of onions costs $1.77. To calculate the price of 5 pounds of onions, a proportion could show:

$$\frac{3}{1.77} = \frac{5}{x}$$

However, by knowing the unit rate, the value of pounds of onions is multiplied by the unit price. The unit price is calculated: $1.77/3lb = \$0.59/lb$. Multiplying the weight of the onions by the unit price yields:

$$5lb \times \frac{\$0.59}{lb} = \$2.95$$

The *lb.* units cancel out.

Similar to unit-rate problems, unit conversions appear in real-world scenarios including cooking, measurement, construction, and currency. Given the conversion rate, unit conversions are written as a fraction (ratio) and multiplied by a quantity in one unit to convert it to the corresponding unit. To determine how many minutes are in $3\frac{1}{2}$ hours, the conversion rate of 60 minutes to 1 hour is written as $\frac{60\ min}{1h}$. Multiplying the quantity by the conversion rate results in:

$$3\frac{1}{2}h \times \frac{60\ min}{1h} = 210\ min$$

(The *h* unit is canceled.) To convert a quantity in minutes to hours, the fraction for the conversion rate is flipped to cancel the *min* unit. To convert 195 minutes to hours, $195min \times \frac{1h}{60\ min}$ is multiplied. The result is $\frac{195h}{60}$ which reduces to $3\frac{1}{4}$h.

Converting units may require more than one multiplication. The key is to set up conversion rates so that units cancel each other out and the desired unit is left. To convert 3.25 yards to inches, given that 1yd = 3ft and 12in = 1ft, the calculation is performed by multiplying:

$$3.25\ yd \times \frac{3ft}{1yd} \times \frac{12in}{1ft}$$

The *yd* and *ft* units will cancel, resulting in 117in.

When working with different systems of measurement, conversion from one unit to another may be necessary. The conversion rate must be known to convert units. One method for converting units is to write and solve a proportion. The arrangement of values in a proportion is extremely important. Suppose that a problem requires converting 20 fluid ounces to cups. To do so, a proportion can be written using the conversion rate of 8fl oz = 1c with *x* representing the missing value. The proportion can be written in any of the following ways:

$$\frac{1}{8} = \frac{x}{20}\left(\frac{c\ for\ conversion}{fl\ oz\ for\ conversion} = \frac{unknown\ c}{fl\ oz\ given}\right); \frac{8}{1} = \frac{20}{x}\left(\frac{fl\ oz\ for\ conversion}{c\ for\ conversion} = \frac{fl\ oz\ given}{unknown\ c}\right);$$

$$\frac{1}{x} = \frac{8}{20}\left(\frac{c\ for\ conversion}{unknown\ c} = \frac{fl\ oz\ for\ conversion}{fl\ oz\ given}\right); \frac{x}{1} = \frac{20}{8}\left(\frac{unknown\ c}{c\ for\ conversion} = \frac{fl\ oz\ given}{fl\ oz\ for\ conversion}\right)$$

To solve a proportion, the ratios are cross-multiplied and the resulting equation is solved. When cross-multiplying, all four proportions above will produce the same equation:

$$(8)(x) = (20)(1) \rightarrow 8x = 20$$

Dividing by 8 to isolate the variable $x$, the result is $x = 2.5$. The variable $x$ represented the unknown number of cups. Therefore, the conclusion is that 20 fluid ounces converts (is equal) to 2.5 cups.

Sometimes converting units requires writing and solving more than one proportion. Suppose an exam question asks to determine how many hours are in 2 weeks. Without knowing the conversion rate between hours and weeks, this can be determined knowing the conversion rates between weeks and days, and between days and hours. First, weeks are converted to days, then days are converted to hours. To convert from weeks to days, the following proportion can be written:

$$\frac{7}{1} = \frac{x}{2} \left( \frac{days\ conversion}{weeks\ conversion} = \frac{days\ unknown}{weeks\ given} \right)$$

Cross-multiplying produces:

$$(7)(2) = (x)(1) \rightarrow 14 = x$$

Therefore, 2 weeks is equal to 14 days. Next, a proportion is written to convert 14 days to hours:

$$\frac{24}{1} = \frac{x}{14} \left( \frac{conversion\ hours}{conversion\ days} = \frac{unknown\ hours}{given\ days} \right)$$

Cross-multiplying produces:

$$(24)(14) = (x)(1) \rightarrow 336 = x$$

Therefore, the answer is that there are 336 hours in 2 weeks.

## Scatterplots

A *scatter plot* is a way to visually represent the relationship between two variables. Each variable has its own axis, and usually the independent variable is plotted on the horizontal axis while the dependent variable is plotted on the vertical axis. Data points are plotted in a process that's similar to how ordered pairs are plotted on an *xy*-plane. Once all points from the data set are plotted, the scatter plot is

finished. Below is an example of a scatter plot that's plotting the quality and price of an item. Note that price is the independent variable and quality is the dependent variable:

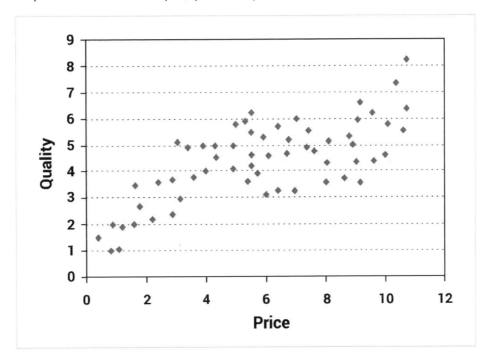

In this example, the quality of the item increases as the price increases.

*Regression lines* are a way to calculate a relationship between the independent variable and the dependent variable. A straight line means that there's a linear trend in the data. Technology can be used to find the equation of this line (e.g., a graphing calculator or Microsoft Excel®). In either case, all of the data points are entered, and a line is "fit" that best represents the shape of the data. If the line of best-fit has a positive slope (rises from left to right), then the variables have a positive correlation. If the line of best-fit has a negative slope (falls from left to right), then the variables have a negative correlation. If a line of best-fit cannot be drawn, then no correlation exists. A positive or negative correlation can be categorized as strong or weak, depending on how closely the points are graphed around the line of best-fit. Other functions used to model data sets include quadratic and exponential models.

Regression lines can be used to estimate data points not already given. Consider a data set with the average daily temperature at the beach and number of beach visitors. If an equation of a line is found that fits this data set, its input is the average daily temperature and its output is the projected number of visitors. Thus, the number of beach visitors on a 100-degree day can be estimated. The output is a data point on the regression line, and the number of daily visitors is expected to be greater than on a 96-degree day because the regression line has a positive slope.

The formula for a regression line is $y = mx + b$, where $m$ is the slope and $b$ is the y-intercept. Both the slope and y-intercept are found in the *Method of Least Squares*, which is the process of finding the equation of the line through minimizing residuals. The slope represents the rate of change in $y$ as $x$ gets larger. Therefore, because $y$ is the dependent variable, the slope actually provides the predicted values given the independent variable. The y-intercept is the predicted value for when the independent variable equals zero. In the temperature example, the y-intercept is the expected number of beach visitors for a very cold average daily temperature of zero degrees.

## Investigating Key Features of a Graph

When a linear equation is written in standard form, $Ax + By = C$, it is easy to identify the $x$- and $y$-intercepts for the graph of the line. Just as the $y$-intercept is the point at which the line intercepts the $y$-axis, the $x$-intercept is the point at which the line intercepts the $x$-axis. At the $y$-intercept, $x = 0$, and at the $x$-intercept, $y = 0$. Given an equation in standard form, substitute $x = 0$ to find the $y$-intercept, and substitute $y = 0$ to find the $x$-intercept. For example, to graph $3x + 2y = 6$, substituting 0 for $y$ results in:

$$3x + 2(0) = 6$$

Solving for $x$ yields $x = 2$; therefore, an ordered pair for the line is (2, 0). Substituting 0 for $x$ results in:

$$3(0) + 2y = 6$$

Solving for $y$ yields $y = 3$; therefore, an ordered pair for the line is (0, 3). Plot the two ordered pairs (the $x$- and $y$-intercepts), and construct a straight line through them.

### T - chart

| x | y |
|---|---|
| 0 | 3 |
| 2 | 0 |

### Intercepts

x - intercept : (2,0)

y - intercept : (0,3)

(0,3)   3x + 2y = 6   (2,0)

The standard form of a quadratic function is:

$$y = ax^2 + bx + c$$

The graph of a quadratic function is a U-shaped (or upside-down U) curve, called a *parabola*, which is symmetric about a vertical line (axis of symmetry). To graph a parabola, determine its vertex (high or low point for the curve) and at least two points on each side of the axis of symmetry.

Given a quadratic function in standard form, the axis of symmetry for its graph is the line:

$$x = -\frac{b}{2a}$$

The vertex for the parabola has an x-coordinate of $-\frac{b}{2a}$. To find the y-coordinate for the vertex, substitute the calculated x-coordinate. To complete the graph, select two different x-values, and substitute them into the quadratic function to obtain the corresponding y-values. This will give two points on the parabola. Use these two points and the axis of symmetry to determine the two points corresponding to these. The corresponding points are the same distance from the axis of symmetry (on the other side) and contain the same y-coordinate.

Plotting the vertex and four other points on the parabola allows for construction of the curve.

## Quadratic Function

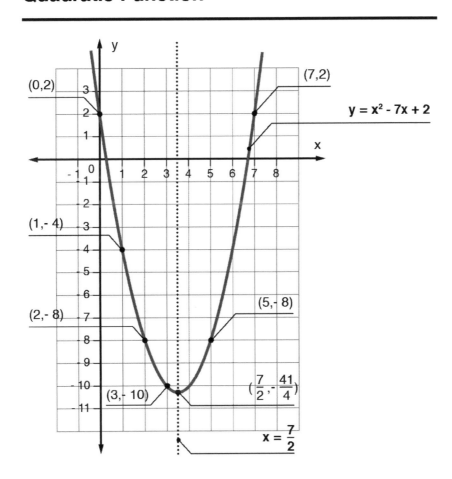

Exponential functions have a general form of $y = (a)(b^x)$. The graph of an exponential function is a curve that slopes upward or downward from left to right. The graph approaches a line, called an *asymptote,* as x or y increases or decreases. To graph the curve for an exponential function, select x-

values, and substitute them into the function to obtain the corresponding *y*-values. A general rule of thumb is to select three negative values, zero, and three positive values.

Plotting the seven points on the graph for an exponential function should allow for the construction of a smooth curve through them.

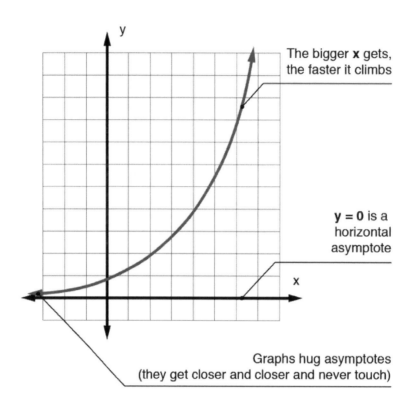

## Comparing Linear and Exponential Growth

Linear functions are simpler than exponential functions, and the independent variable $x$ has an exponent of 1. Written in the most common form, $y = mx + b$, the coefficient of $x$ tells how fast the function grows at a constant rate, and the $b$-value tells the starting point. An exponential function has an independent variable in the exponent $y = ab^x$. The graph of these types of functions is described as *growth* or *decay*, based on whether the base, $b$, is greater than or less than 1. These functions are different from quadratic functions because the base stays constant. A common base is base $e$.

The following two functions model a linear and exponential function respectively: $y = 2x$ and $y = 2^x$. Their graphs are shown below. The first graph, modeling the linear function, shows that the growth is constant over each interval. With a horizontal change of 1, the vertical change is 2. It models a constant positive growth. The second graph models the exponential function, where the horizontal change of 1 yields a vertical change that increases more and more. The exponential graph gets very close to the $x$-

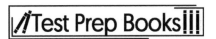

axis, but never touches it, meaning there is an asymptote there. The y-value can never be zero because the base of 2 can never be raised to an input value that yields an output of zero.

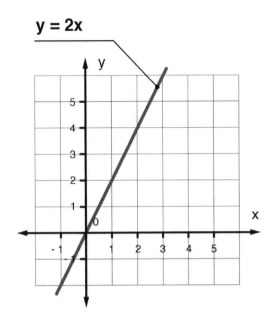

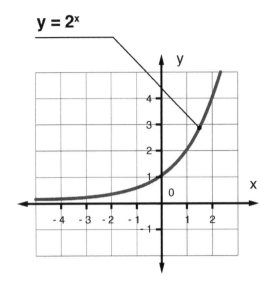

Given a table of values, the type of function can be determined by observing the change in $y$ over equal intervals. For example, the tables below model two functions. The changes in interval for the $x$-values is 1 for both tables. For the first table, the $y$-values increase by 5 for each interval. Since the change is constant, the situation can be described as a linear function. The equation would be:

$$y = 5x + 3$$

For the second table, the change for $y$ is 5, 20, 100, and 500, respectively. The increases are multiples of 5, meaning the situation can be modeled by an exponential function. The equation $y = 5^x + 3$ models this situation.

| x | y |
|---|---|
| 0 | 3 |
| 1 | 8 |
| 2 | 13 |
| 3 | 18 |
| 4 | 23 |

| x | y |
|---|---|
| 0 | 3 |
| 1 | 8 |
| 2 | 28 |
| 3 | 128 |
| 4 | 628 |

## Two-Way Tables

Data that isn't described using numbers is known as *categorical data.* For example, age is numerical data but hair color is categorical data. Categorical data is summarized using two-way frequency tables. A *two-way frequency table* counts the relationship between two sets of categorical data. There are rows and

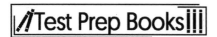

columns for each category, and each cell represents frequency information that shows the actual data count between each combination.

For example, the graphic on the left-side below is a two-way frequency table showing the number of girls and boys taking language classes in school. Entries in the middle of the table are known as the *joint frequencies*. For example, the number of girls taking French class is 12, which is a joint frequency. The totals are the *marginal frequencies*. For example, the total number of boys is 20, which is a marginal frequency. If the frequencies are changed into percentages based on totals, the table is known as a *two-way relative frequency table*. Percentages can be calculated using the table total, the row totals, or the column totals.

Here's the process of obtaining the two-way relative frequency table using the table total:

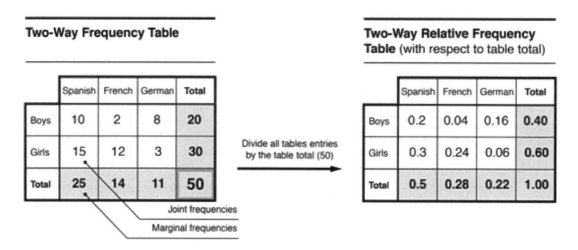

The middle entries are known as *joint probabilities* and the totals are *marginal probabilities*. In this data set, it appears that more girls than boys take Spanish class. However, that might not be the case because more girls than boys were surveyed and the results might be misleading. To avoid such errors, *conditional relative frequencies* are used. The relative frequencies are calculated based on a row or column.

Here are the conditional relative frequencies using column totals:

**Two-Way Frequency Table**

| | Spanish | French | German | Total |
|---|---|---|---|---|
| Boys | 10 | 2 | 8 | 20 |
| Girls | 15 | 12 | 3 | 30 |
| Total | 25 | 14 | 11 | 50 |

Divide each column entry by that column's total

**Two-Way Relative Frequency Table** (with respect to table total)

| | Spanish | French | German | Total |
|---|---|---|---|---|
| Boys | 0.4 | 0.14 | 0.73 | 0.4 |
| Girls | 0.6 | 0.86 | 0.27 | 0.6 |
| Total | 1.00 | 1.00 | 1.00 | 1.00 |

Two-way frequency tables can help in making many conclusions about the data. If either the row or column of conditional relative frequencies differs between each row or column of the table, then an association exists between the two categories. For example, in the above tables, the majority of boys are taking German while the majority of girls are taking French. If the frequencies are equal across the rows, there is no association and the variables are labelled as independent. It's important to note that the association does exist in the above scenario, though these results may not occur the next semester when students are surveyed.

When measuring event probabilities, two-way frequency tables can be used to report the raw data and then used to calculate probabilities. If the frequency tables are translated into relative frequency tables, the probabilities presented in the table can be plugged directly into the formulas for conditional probabilities. By plugging in the correct frequencies, the data from the table can be used to determine if events are independent or dependent.

*Conditional probability* is the probability that event A will happen given that event B has already occurred. An example of this is calculating the probability that a person will eat dessert once they have eaten dinner. This is different than calculating the probability of a person just eating dessert.

The formula for the conditional probability of event A occurring given B is:

$$P(A|B) = \frac{P\ (A \text{ and } B)}{P(B)}$$

It's defined to be the probability of both A and B occurring divided by the probability of event B occurring. If A and B are independent, then the probability of both A and B occurring is equal to $P(A)P(B)$, so $P(A|B)$ reduces to just $P(A)$. This means that A and B have no relationship, and the probability of A occurring is the same as the conditional probability of A occurring given B. Similarly:

$$P(B|A) = \frac{P\ (B \text{ and } A\ )}{P(A)} = P(B)$$

(if A and B are independent)

Two events aren't always independent. For examples, females with glasses and brown hair aren't independent characteristics. There definitely can be overlap because females with brown hair can wear glasses. Also, two events that exist at the same time don't have to have a relationship. For example, even if all females in a given sample are wearing glasses, the characteristics aren't related. In this case, the probability of a brunette wearing glasses is equal to the probability of a female being a brunette multiplied by the probability of a female wearing glasses. This mathematical test of $P(A \cap B) = P(A)P(B)$ verifies that two events are independent.

Conditional probability is the probability that an event occurs given that another event has happened. If the two events are related, the probability that the second event will occur changes if the other event has happened. However, if the two events aren't related and are therefore independent, the first event to occur won't impact the probability of the second event occurring.

## Making Inferences About Population Parameters

*Inferential statistics* attempts to use data about a subset of some population to make inferences about the rest of the population. An example of this would be taking a collection of students who received tutoring and comparing their results to a collection of students who did not receive tutoring, then using that comparison to try to predict whether the tutoring program in question is beneficial.

To be sure that inferences have a high probability of being true for the whole population, the subset that is analyzed needs to resemble a miniature version of the population as closely as possible. For this reason, statisticians like to choose random samples from the population to study, rather than picking a specific group of people based on some similarity. For example, studying the incomes of people who live in Portland does not tell anything useful about the incomes of people who live in Tallahassee.

A *population* is the entire set of people or things of interest. Suppose a study is intended to determine the number of hours of sleep per night for college females in the United States. The population would consist of EVERY college female in the country. A *sample* is a subset of the population that may be used for the study. It would not be practical to survey every female college student, so a sample might consist of one hundred students per school from twenty different colleges in the country. From the results of the survey, a sample statistic can be calculated. A sample statistic is a numerical characteristic of the sample data, including mean and variance. A sample statistic can be used to estimate a corresponding population parameter. A population parameter is a numerical characteristic of the entire population. Suppose our sample data had a mean (average) of 5.5. This sample statistic can be used as an estimate of the population parameter (average hours of sleep for every college female in the United States).

A *population parameter* is usually unknown and therefore estimated using a sample statistic. This estimate may be very accurate or relatively inaccurate based on errors in sampling. A *confidence interval* indicates a range of values likely to include the true population parameter. These are constructed at a given confidence level, such as 95 percent. This means that if the same population is sampled repeatedly, the true population parameter would occur within the interval for 95 percent of the samples.

The accuracy of a population parameter based on a sample statistic may also be affected by *measurement error*. Measurement error is the difference between a quantity's true value and its measured value. Measurement error can be divided into random error and systematic error. An example of random error for the previous scenario would be a student reporting 8 hours of sleep when she sleeps 7 hours per night. Systematic errors are those attributed to the measurement system. Suppose the sleep survey gave response options of 2, 4, 6, 8, or 10 hours. This would lead to systematic measurement error.

## Statistics

The field of statistics describes relationships between quantities that are related, but not necessarily in a deterministic manner. For example, a graduating student's salary will often be higher when the student graduates with a higher GPA, but this is not always the case. Likewise, people who smoke tobacco are more likely to develop lung cancer, but, in fact, it is possible for non-smokers to develop the disease as well. *Statistics* describes these kinds of situations, where the likelihood of some outcome depends on the starting data.

Comparing data sets within statistics can mean many things. The first way to compare data sets is by looking at the center and spread of each set. The center of a data set is measured by mean, median, and mode.

Suppose that $X$ is a set of data points $(x_1, x_2, x_3, \ldots x_n)$ and some description of the general properties of this data need to be found.

The first property that can be defined for this set of data is the *mean*. To find the mean, add up all the data points, then divide by the total number of data points. This can be expressed using *summation notation* as:

$$\bar{X} = \frac{x_1 + x_2 + x_3 + \cdots + x_n}{n} = \frac{1}{n} \sum_{i=1}^{n} x_i$$

For example, suppose that in a class of 10 students, the scores on a test were 50, 60, 65, 65, 75, 80, 85, 85, 90, 100. Therefore, the average test score will be:

$$\frac{1}{10}(50 + 60 + 65 + 65 + 75 + 80 + 85 + 85 + 90 + 100) = 75.5$$

The mean is a useful number if the distribution of data is normal (more on this later), which roughly means that the frequency of different outcomes has a single peak and is roughly equally distributed on both sides of that peak. However, it is less useful in some cases where the data might be split or where there are some *outliers*. Outliers are data points that are far from the rest of the data. For example, suppose there are 90 employees and 10 executives at a company. The executives make $1000 per hour, and the employees make $10 per hour. Therefore, the average pay rate will be $\frac{1000 \cdot 10 + 10 \cdot 90}{100} = 109$, or $109 per hour. In this case, this average is not very descriptive.

Another useful measurement is the *median*. In a data set $X$ consisting of data points $x_1, x_2, x_3, \ldots x_n$, the median is the point in the middle. The middle refers to the point where half the data comes before it and half comes after, when the data is recorded in numerical order. If $n$ is odd, then the median is:

$$x_{\frac{n+1}{2}}$$

If $n$ is even, it is defined as $\frac{1}{2}\left(x_{\frac{n}{2}} + x_{\frac{n}{2}+1}\right)$, the mean of the two data points closest to the middle of the data points. In the previous example of test scores, the two middle points are 75 and 80. Since there is no single point, the average of these two scores needs to be found. The average is $\frac{75+80}{2} = 77.5$. The median is generally a good value to use if there are a few outliers in the data. It prevents those outliers from affecting the "middle" value as much as when using the mean.

Since an outlier is a data point that is far from most of the other data points in a data set, this means an outlier also is any point that is far from the median of the data set. The outliers can have a substantial effect on the mean of a data set, but they usually do not change the median or mode, or do not change them by a large quantity. For example, consider the data set (3, 5, 6, 6, 6, 8). This has a median of 6 and a mode of 6, with a mean of $\frac{34}{6} \approx 5.67$. Now, suppose a new data point of 1000 is added so that the data set is now (3, 5, 6, 6, 6, 8, 1000). This does not change the median or mode, which are both still 6. However, the average is now $\frac{1034}{7}$, which is approximately 147.7. In this case, the median and mode will be better descriptions for most of the data points.

The reason for outliers in a given data set is a complicated problem. It is sometimes the result of an error by the experimenter, but often they are perfectly valid data points that must be taken into consideration.

Test Prep Books

One additional measure to define for X is the *mode*. This is the data point that appears more frequently. If two or more data points all tie for the most frequent appearance, then each of them is considered a mode. In the case of the test scores, where the numbers were 50, 60, 65, 65, 75, 80, 85, 85, 90, 100, there are two modes: 65 and 85.

The spread of a data set refers to how far the data points are from the center (mean or median). The spread can be measured by the range or the quartiles and interquartile range. A data set with data points clustered around the center will have a small spread. A data set covering a wide range will have a large spread. The *interquartile range (IQR)* is the range of the middle 50 percent of the data set. This range can be seen in the large rectangle on a box plot. The *standard deviation (σ)* quantifies the amount of variation with respect to the mean. A lower standard deviation shows that the data set doesn't differ greatly from the mean. A larger standard deviation shows that the data set is spread out farther from the mean. The formula for standard deviation is:

$$\sigma = \sqrt{\frac{\sum(x - \bar{x})^2}{n - 1}}$$

$x$ is each value in the data set, $\bar{x}$ is the mean, and $n$ is the total number of data points in the set.

The shape of a data set is another way to compare two or more sets of data. If a data set isn't symmetric around its mean, it's said to be *skewed*. If the tail to the left of the mean is longer, it's said to be *skewed to the left*. In this case, the mean is less than the median. Conversely, if the tail to the right of the mean is longer, it's said to be *skewed to the right* and the mean is greater than the median. When classifying a data set according to its shape, its overall *skewness* is being discussed. If the mean and median are equal, the data set isn't skewed; it is *symmetric*.

## Evaluating Reports

The presentation of statistics can be manipulated to produce a desired outcome. Consider the statement, "Four out of five dentists recommend our toothpaste." This is a vague statement that is obviously biased. (Who are the five dentists this statement references?) This statement is very different from the statement, "Four out of every five dentists recommend our toothpaste." Whether intentional or unintentional, statistics can be misleading. Statistical reports should be examined to verify the validity and significance of the results. The context of the numerical values allows for deciphering the meaning, intent, and significance of the survey or study. Questions on this material will require students to use critical-thinking skills to justify or reject results and conclusions.

When analyzing a report, consider who conducted the study and their intent. Was it performed by a neutral party or by a person or group with a vested interest? A study on health risks of smoking performed by a health insurance company would have a much different intent than one performed by a cigarette company. Consider the sampling method and the data collection method. Was it a true random sample of the population, or was one subgroup overrepresented or underrepresented?

The three most common types of data gathering techniques are sample surveys, experiments, and observational studies. *Sample surveys* involve collecting data from a random sample of people from a desired population. The measurement of the variable is only performed on this set of people. To have accurate data, the sampling must be unbiased and random. For example, surveying students in an advanced calculus class on how much they enjoy math classes is not a useful sample if the population should be all college students based on the research question. An *experiment* is the method in which a

hypothesis is tested using a trial-and-error process. A cause and the effect of that cause are measured, and the hypothesis is accepted or rejected. Experiments are usually completed in a controlled environment where the results of a control population are compared to the results of a test population. The groups are selected using a randomization process in which each group has a representative mix of the population being tested. Finally, an *observational study* is similar to an experiment. However, this design is used when there cannot be a designed control and test population because of circumstances (e.g., lack of funding or unrealistic expectations). Instead, existing control and test populations must be used, so this method has a lack of randomization.

Consider the sleep study scenario from the previous section. If all twenty schools included in the study were state colleges, the results may be biased due to a lack of private-school participants. Consider the measurement system used to obtain the data. Was the system accurate and precise, or was it a flawed system? If, for the sleep study, the possible responses were limited to 2, 4, 6, 8, or 10 hours, it could be argued that the measurement system was flawed. Would odd numbers be rounded up or down? Without clarity of the system, the results could vary greatly. What about students who sleep 12 hours per night? The closest option for them would be 10 hours, which is significantly less.

Every scenario involving statistical reports will be different. The key is to examine all aspects of the study before determining whether to accept or reject the results and corresponding conclusions.

## *Passport to Advanced Math*

### Creating Quadratic and Exponential Functions

A polynomial of degree 2 is called *quadratic*. Every quadratic function can be written in the form:

$$ax^2 + bx + c$$

The graph of a quadratic function, $y = ax^2 + bx + c$, is called a *parabola*. Parabolas are vaguely U-shaped.

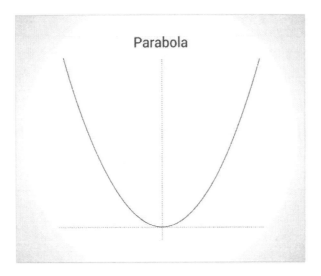
Parabola

Whether the parabola opens upward or downward depends on the sign of *a*. If *a* is positive, then the parabola will open upward. If *a* is negative, then the parabola will open downward. The value of *a* will also affect how wide the parabola is. If the absolute value of *a* is large, then the parabola will be fairly skinny. If the absolute value of *a* is small, then the parabola will be quite wide.

Changes to the value of *b* affect the parabola in different ways, depending on the sign of *a*. For positive values of *a*, increasing *b* will move the parabola to the left, and decreasing *b* will move the parabola to the right. On the other hand, if *a* is negative, the effects will be the opposite: increasing *b* will move the parabola to the right, while decreasing *b* will move the parabola to the left.

Changes to the value of *c* move the parabola vertically. The larger that *c* is, the higher the parabola gets. This does not depend on the value of *a*.

The quantity $D = b^2 - 4ac$ is called the *discriminant* of the parabola. When the discriminant is positive, then the parabola has two real zeros, or *x* intercepts. However, if the discriminant is negative, then there are no real zeros, and the parabola will not cross the *x*-axis. The highest or lowest point of the parabola is called the *vertex*. If the discriminant is zero, then the parabola's highest or lowest point is on the *x*-axis, and it will have a single real zero. The x-coordinate of the vertex can be found using the equation $x = -\frac{b}{2a}$. Plug this x-value into the equation and find the y-coordinate.

A quadratic equation is often used to model the path of an object thrown into the air. The x-value can represent the time in the air, while the y-value can represent the height of the object. In this case, the maximum height of the object would be the y-value found when the x-value is $-\frac{b}{2a}$.

An *exponential function* is a function of the form $f(x) = b^x$, where *b* is a positive real number other than 1. In such a function, *b* is called the *base*.

The *domain* of an exponential function is all real numbers, and the *range* is all positive real numbers. There will always be a horizontal asymptote of $y = 0$ on one side. If *b* is greater than 1, then the graph will be increasing moving to the right. If *b* is less than 1, then the graph will be decreasing moving to the right. Exponential functions are one-to-one. The basic exponential function graph will go through the point (0,1).

## Example
Solve $5^{x+1} = 25$.

Get the x out of the exponent by rewriting the equation $5^{x+1} = 5^2$ so that both sides have a base of 5.

Since the bases are the same, the exponents must be equal to each other.

This leaves $x + 1 = 2$ or $x = 1$.

To check the answer, the x-value of 1 can be substituted back into the original equation.

## Determining Forms of Expressions

There is a four-step process in problem-solving that can be used as a guide:

1. Understand the problem and determine the unknown information.

2. Translate the verbal problem into an algebraic equation.

3. Solve the equation by using inverse operations.

4. Check the work and answer the given question.

## Example

Three times the sum of a number plus 4 equals the number plus 8. What is the number?

The first step is to determine the unknown, which is the number, or $x$.

The second step is to translate the problem into the equation, which is:

$$3(x + 4) = x + 8$$

The equation can be solved as follows:

| | |
|---|---|
| $3x + 12 = x + 8$ | Apply the distributive property |
| $3x = x - 4$ | Subtract 12 from both sides of the equation |
| $2x = -4$ | Subtract x from both sides of the equation |
| $x = -2$ | Divide both sides of the equation by 2 |

The final step is checking the solution. Plugging the value for x back into the equation yields the following problem:

$$3(-2) + 12 = -2 + 8$$

Using the order of operations shows that a true statement is made: $6 = 6$

The four-step process of problem solving can be used with geometric reasoning problems as well. There are many geometric properties and terminology included within geometric reasoning.

For example, the perimeter of a rectangle can be written in the terms of the width, or the width can be written in terms of the length.

## Example

The width of a rectangle is 2 centimeters less than the length. If the perimeter of the rectangle is 44 centimeters, then what are the dimensions of a rectangle?

The first step is to determine the unknown, which is in terms of the length, $l$.

The second step is to translate the problem into the equation using the perimeter of a rectangle,

$$P = 2l + 2w$$

The width is the length minus 2 centimeters. The resulting equation is:

$$2l + 2(l - 2) = 44$$

The equation can be solved as follows:

| | |
|---|---|
| $2l + 2l - 4 = 44$ | Apply the distributive property on the left side of the equation |
| $4l - 4 = 44$ | Combine like terms on the left side of the equation |
| $4l = 48$ | Add 4 to both sides of the equation |
| $l = 12$ | Divide both sides of the equation by 4 |

The length of the rectangle is 12 centimeters. The width is the length minus 2 centimeters, which is 10 centimeters. Checking the answers for length and width forms the following equation:

$$44 = 2(12) + 2(10)$$

The equation can be solved using the order of operations to form a true statement: $44 = 44$.

Equations can also be created from complementary angles (angles that add up to 90°) and supplementary angles (angles that add up to 180°).

Example
Two angles are complementary. If one angle is four times the other angle, what is the measure of each angle?

The first step is to determine the unknown, which is the measure of the angle.

The second step is to translate the problem into the equation using the known statement: the sum of two complementary angles is 90°. The resulting equation is:

$$4x + x = 90$$

The equation can be solved as follows:

| $5x = 90$ | Combine like terms on the left side of the equation |
|---|---|
| $x = 18$ | Divide both sides of the equation by 5 |

The first angle is 18° and the second angle is 4 times the unknown, which is 4 times 18 or 72°.

Going back to check the answer with the original question, 72 and 18 have a sum of 90, making them complementary angles. Seventy-two degrees is also four times the other angle, 18 degrees.

## Creating Equivalent Expressions Involving Exponents and Radicals

An *exponent* is written as $a^b$. In this expression, $a$ is called the *base* and $b$ is called the *exponent*. It is properly stated that $a$ is raised to the *n*-th power. Therefore, in the expression $2^3$, the exponent is 3, while the base is 2. Such an expression is called an *exponential expression*. Note that when the exponent is 2, it is called *squaring* the base, and when it is 3, it is called *cubing* the base.

When the exponent is a positive integer, this indicates the base is multiplied by itself the number of times written in the exponent. So, in the expression $2^3$, multiply 2 by itself with 3 copies of 2:

$$2^3 = 2 \times 2 \times 2 = 8$$

One thing to notice is that, for positive integers $n$ and $m$, $a^n a^m = a^{n+m}$ is a rule. In order to make this rule be true for an integer of 0, $a^0 = 1$, so that $a^n a^0 = a^{n+0} = a^n$. And, in order to make this rule be true for negative exponents, $a^{-n} = \frac{1}{a^n}$.

Another rule for simplifying expressions with exponents is shown by the following equation: $(a^m)^n = a^{mn}$. This is true for fractional exponents as well. So, for a positive integer, define $a^{\frac{1}{n}}$ to be the number that, when raised to the *n*-th power, provides $a$. In other words, $(a^{\frac{1}{n}})^n = a$ is the desired equation. It

40

should be noted that $a^{\frac{1}{n}}$ is the *n*-th root of *a*. This also can be written as $a^{\frac{1}{n}} = \sqrt[n]{a}$. The symbol on the right-hand side of this equation is called a *radical*. If the root is left out, assume that the 2$^{nd}$ root should be taken, also called the *square* root: $a^{\frac{1}{2}} = \sqrt[2]{a} = \sqrt{a}$. Additionally, $\sqrt[3]{a}$ is also called the *cube* root.

Note that when multiple roots exist, $a^{\frac{1}{n}}$ is defined to be the *positive* root. So, $4^{\frac{1}{2}} = 2$. Also note that negative numbers do not have even roots in the real numbers.

This also enables finding exponents for any rational number:

$$a^{\frac{m}{n}} = (a^{\frac{1}{n}})^m = (a^m)^{\frac{1}{n}}$$

In fact, the exponent can be any real number. In general, the following rules for exponents should be used for any numbers $a, b, m,$ and $n$.

- $a^1 = a$.
- $1^a = 1$.
- $a^0 = 1$.
- $a^m a^n = a^{m+n}$.
- $\frac{a^m}{a^n} = a^{m-n}$
- $(a^m)^n = a^{m \times n}$.
- $(ab)^m = a^m b^m$.
- $(\frac{a}{b})^m = \frac{a^m}{b^m}$.

As an example of applying these rules, consider the problem of simplifying the expression $(3x^2y)^3(2xy^4)$. Start by simplifying the left term using the sixth rule listed. Applying this rule yields the following expression: $27x^6y^3(2xy^4)$. The exponents can now be combined with base *x* and the exponents with base *y*. Multiply the coefficients to yield $54x^7y^7$.

In mathematical expressions containing exponents and other operations, the order of operations must be followed. PEMDAS states that exponents are calculated after any parenthesis and grouping symbols but before any multiplication, division, addition, and subtraction.

Here are some of the most important properties of exponents and roots: if *n* is an integer, and if $a^n = b^n$, then $a = b$ if *n* is odd; but $a = \pm b$ if *n* is even. Similarly, if the roots of two things are equal, $\sqrt[n]{a} = \sqrt[n]{b}$, then $a = b$. This means that when starting with a true equation, both sides of that equation can be raised to a given power to obtain another true equation. Beware that when an even-powered root is taken on both sides of the equation, a $\pm$ in the result. For example, given the equation $x^2 = 16$, take the square root of both sides to solve for x. This results in the answer $x = \pm 4$ because $(-4)^2 = 16$ and $(4)^2 = 16$.

Another property is that if $a^n = a^m$, then $n = m$. This is true for any real numbers *n* and *m*.

For solving the equation $\sqrt{x + 2} - 1 = 3$, start by moving the -1 over to the right-hand side. This is performed by adding 1 to both sides, which yields:

$$\sqrt{x + 2} = 4$$

41

Now, square both sides, but remember that by squaring both sides, the signs are irrelevant. This yields $x + 2 = 16$, which simplifies to give $x = 14$.

Now consider the problem $(x + 1)^4 = 16$. To solve this, take the $4^{th}$ root of both sides, which means an ambiguity in the sign will be introduced because it is an even root:

$$\sqrt[4]{(x + 1)^4} = \pm\sqrt[4]{16}$$

The right-hand side is 2, since $2^4 = 16$. Therefore, $x + 1 = \pm 2$ or $x = -1 \pm 2$. Thus, the two possible solutions are $x = -3$ and $x = 1$.

Remember that when solving equations, the answer can be checked by plugging the solution back into the problem to make a true statement.

## Creating Equivalent Forms of Expressions

*Algebraic expressions* are made up of numbers, variables, and combinations of the two, using mathematical operations. Expressions can be rewritten based on their factors. For example, the expression $6x + 4$ can be rewritten as $2(3x + 2)$ because 2 is a factor of both $6x$ and 4. More complex expressions can also be rewritten based on their factors. The expression $x^4 - 16$ can be rewritten as $(x^2 - 4)(x^2 + 4)$. This is a different type of factoring, where a difference of squares is factored into a sum and difference of the same two terms. With some expressions, the factoring process is simple and only leads to a different way to represent the expression. With others, factoring and rewriting the expression leads to more information about the given problem.

In the following quadratic equation, factoring the binomial leads to finding the zeros of the function:

$$x^2 - 5x + 6 = y$$

This equations factors into $(x - 3)(x - 2) = y$, where 2 and 3 are found to be the zeros of the function when y is set equal to zero. The zeros of any function are the x-values where the graph of the function on the coordinate plane crosses the x-axis.

Factoring an equation is a simple way to rewrite the equation and find the zeros, but factoring is not possible for every quadratic. Completing the square is one way to find zeros when factoring is not an option. The following equation cannot be factored:

$$x^2 + 10x - 9 = 0$$

The first step in this method is to move the constant to the right side of the equation, making it:

$$x^2 + 10x = 9$$

Then, the coefficient of x is divided by 2 and squared. This number is then added to both sides of the equation, to make the equation still true. For this example, $\left(\frac{10}{2}\right)^2 = 25$ is added to both sides of the equation to obtain:

$$x^2 + 10x + 25 = 9 + 25$$

This expression simplifies to $x^2 + 10x + 25 = 34$, which can then be factored into:

$$(x + 5)^2 = 34$$

Solving for x then involves taking the square root of both sides and subtracting 5. This leads to two zeros of the function:

$$x = \pm\sqrt{34} - 5$$

Depending on the type of answer the question seeks, a calculator may be used to find exact numbers.

Given a quadratic equation in standard form— $ax^2 + bx + c = 0$ —the sign of $a$ tells whether the function has a minimum value or a maximum value. If $a > 0$, the graph opens up and has a minimum value. If $a < 0$, the graph opens down and has a maximum value. Depending on the way the quadratic equation is written, multiplication may need to occur before a max/min value is determined.

There are also properties of numbers that are true for certain operations. The *commutative* property allows the order of the terms in an expression to change while keeping the same final answer. Both addition and multiplication can be completed in any order and still obtain the same result. However, order does matter in subtraction and division. The *associative* property allows any terms to be "associated" by parenthesis and retain the same final answer. For example,

$$(4 + 3) + 5 = 4 + (3 + 5)$$

Both addition and multiplication are associative; however, subtraction and division do not hold this property. The *distributive* property states that $a(b + c) = ab + ac$. It is a property that involves both addition and multiplication, and the $a$ is distributed onto each term inside the parentheses.

The expression $4(3 + 2)$ is simplified using the order of operations. Simplifying inside the parenthesis first produces $4 \times 5$, which equals 20. The expression $4(3 + 2)$ can also be simplified using the distributive property:

$$4(3 + 2) = 4 \times 3 + 4 \times 2 = 12 + 8 = 20$$

Consider the following example: $4(3x - 2)$. The expression cannot be simplified inside the parenthesis because $3x$ and -2 are not like terms and therefore cannot be combined. However, the expression can be simplified by using the distributive property and multiplying each term inside of the parenthesis by the term outside of the parenthesis: $12x - 8$. The resulting equivalent expression contains no like terms, so it cannot be further simplified.

Consider the expression:

$$(3x + 2y + 1) - (5x - 3) + 2(3y + 4)$$

Again, there are no like terms, but the distributive property is used to simplify the expression. Note there is an implied one in front of the first set of parentheses and an implied -1 in front of the second set of parentheses. Distributing the 1, -1, and 2 produces:

$$1(3x) + 1(2y) + 1(1) - 1(5x) - 1(-3) + 2(3y) + 2(4)$$

$$3x + 2y + 1 - 5x + 3 + 6y + 8$$

This expression contains like terms that are combined to produce the simplified expression:

$$-2x + 8y + 12$$

Algebraic expressions are tested to be equivalent by choosing values for the variables and evaluating both expressions. For example, $4(3x - 2)$ and $12x - 8$ are tested by substituting 3 for the variable $x$ and calculating to determine if equivalent values result.

## Solving Quadratic Equations

A *quadratic equation* is an equation in the form:

$$ax^2 + bx + c = 0$$

There are several methods to solve such equations. The easiest method will depend on the quadratic equation in question.

It sometimes is possible to solve quadratic equations by manually *factoring* them. This means rewriting them in the form:

$$(x + A)(x + B) = 0$$

If this is done, then they can be solved by remembering that when $ab = 0$, either $a$ or $b$ must be equal to zero. Therefore, to have $(x + A)(x + B) = 0$, $(x + A) = 0$ or $(x + B) = 0$ is needed. These equations have the solutions $x = -A$ and $x = -B$, respectively.

In order to factor a quadratic equation, note that:

$$(x + A)(x + B) = x^2 + (A + B)x + AB$$

So, if an equation is in the form $x^2 + bx + c$, two numbers, $A$ and $B$, need to be found that will add up to give us $b$, and multiply together to give us $c$.

As an example, consider solving the equation:

$$-3x^2 + 6x + 9 = 0$$

Start by dividing both sides by $-3$, leaving:

$$x^2 - 2x - 3 = 0$$

Now, notice that $1 - 3 = -2$, and also that $(1)(-3) = -3$. This means the equation can be factored into:

$$(x + 1)(x - 3) = 0$$

Now, solve $(x + 1) = 0$ and $(x - 3) = 0$ to get $x = -1$ and $x = 3$ as the solutions.

It is useful when trying to factor to remember that:

$$x^2 + 2xy + y^2$$
$$(x + y)^2, x^2 - 2xy + y^2$$
$$(x - y)^2$$
$$x^2 - y^2$$
$$(x + y)(x - y)$$

However, factoring by hand is often hard to do. If there are no obvious ways to factor the quadratic equation, solutions can still be found by using the *quadratic formula*.

The quadratic formula is:

$$x = \frac{-b \pm \sqrt{b^2 - 4ac}}{2a}$$

This method will always work, although it sometimes can take longer than factoring by hand, if the factors are easy to guess. Using the standard form $ax^2 + bx + c = 0$, plug the values of $a$, $b$, and $c$ from the equation into the formula and solve for x. There will either be two answers, one answer, or no real answer. No real answer comes when the value of the discriminant, the number under the square root, is a negative number. Since there are no real numbers that square to get a negative, the answer will be no real roots.

Here is an example of solving a quadratic equation using the quadratic formula. Suppose the equation to solve is:

$$-2x^2 + 3x + 1 = 0$$

There is no obvious way to factor this, so the quadratic formula is used, with $a = -2, b = 3, c = 1$. After substituting these values into the quadratic formula, it yields this:

$$x = \frac{-3 \pm \sqrt{3^2 - 4(-2)(1)}}{2(-2)}$$

This can be simplified to obtain:

$$\frac{3 \pm \sqrt{9 + 8}}{4}$$

or

$$\frac{3 \pm \sqrt{17}}{4}$$

Challenges can be encountered when asked to find a quadratic equation with specific roots. Given roots $A$ and $B$, a quadratic function can be constructed with those roots by taking $(x - A)(x - B)$. So, in constructing a quadratic equation with roots $x = -2, 3$, it would result in:

$$(x + 2)(x - 3) = x^2 - x - 6$$

Multiplying this by a constant also could be done without changing the roots.

## Adding, Subtracting, and Multiplying Polynomial Expressions

An expression of the form $ax^n$, where $n$ is a non-negative integer, is called a *monomial* because it contains one term. A sum of monomials is called a *polynomial*. For example, $-4x^3 + x$ is a polynomial, while $5x^7$ is a monomial. A function equal to a polynomial is called a *polynomial function*.

The monomials in a polynomial are also called the *terms* of the polynomial.

The constants that precede the variables are called *coefficients*.

The highest value of the exponent of $x$ in a polynomial is called the *degree* of the polynomial. So, $-4x^3 + x$ has a degree of 3, while $-2x^5 + x^3 + 4x + 1$ has a degree of 5. When multiplying polynomials, the degree of the result will be the sum of the degrees of the two polynomials being multiplied.

To add polynomials, add the coefficients of like powers of $x$. For example:

$$(-2x^5 + x^3 + 4x + 1) +$$

$$(-4x^3 + x) - 2x^5 + (1 - 4)x^3 + (4 + 1)x + 1$$

$$-2x^5 - 3x^3 + 5x + 1$$

Likewise, subtraction of polynomials is performed by subtracting coefficients of like powers of $x$. So:

$$(-2x^5 + x^3 + 4x + 1) - (-4x^3 + x)$$

$$-2x^5 + (1 + 4)x^3 + (4 - 1)x + 1$$

$$-2x^5 + 5x^3 + 3x + 1$$

To multiply two polynomials, multiply each term of the first polynomial by each term of the second polynomial and add the results. For example:

$$(4x^2 + x)(-x^3 + x)$$

$$4x^2(-x^3) + 4x^2(x) + x(-x^3) + x(x)$$

$$-4x^5 + 4x^3 - x^4 + x^2$$

In the case where each polynomial has two terms, like in this example, some students find it helpful to remember this as multiplying the First terms, then the Outer terms, then the Inner terms, and finally the Last terms, with the mnemonic FOIL. For longer polynomials, the multiplication process is the same, but there will be, of course, more terms, and there is no common mnemonic to remember each combination.

## Solving Equations with Radicals or Variables in the Denominator

When solving radical and rational equations, extraneous solutions must be accounted for when finding the answers. For example, the equation $\frac{x}{x-5} = \frac{3x}{x+3}$ has two values that create a 0 denominator: $x \neq 5, -3$. When solving for $x$, these values must be considered because they cannot be solutions. In the given equation, solving for $x$ can be done using cross-multiplication, yielding the equation:

$$x(x + 3) = 3x(x - 5)$$

Distributing results in the quadratic equation yields $x^2 + 3x = 3x^2 - 15x$; therefore, all terms must be moved to one side of the equals sign. This results in $2x^2 - 18x = 0$, which in factored form is:

$$2x(x - 9) = 0$$

Setting each factor equal to zero, the apparent solutions are $x = 0$ and $x = 9$. These two solutions are neither 5 nor -3, so they are viable solutions. Neither 0 nor 9 create a 0 denominator in the original equation.

A similar process exists when solving radical equations. One must check to make sure the solutions are defined in the original equations. Solving an equation containing a square root involves isolating the root and then squaring both sides of the equals sign. Solving a cube root equation involves isolating the radical and then cubing both sides. In either case, the variable can then be solved for because there are no longer radicals in the equation.

For example, the following expression is a radical that can be simplified: $\sqrt{24x^2}$. First, the number must be factored out to the highest perfect square. Any perfect square can be taken out of a radical. Twenty-four can be factored into 4 and 6, and 4 can be taken out of the radical. $\sqrt{4} = 2$ can be taken out, and 6 stays underneath. If $x > 0$, $x$ can be taken out of the radical because it is a perfect square. The simplified radical is $2x\sqrt{6}$. An approximation can be found using a calculator.

## Solving a System with One Linear Equation and One Quadratic Equation

A system of equations may also be made up of a linear and a quadratic equation. These systems may have one solution, two solutions, or no solutions. The graph of these systems involves one straight line and one parabola. Algebraically, these systems can be solved by solving the linear equation for one variable and plugging that answer in to the quadratic equation. If possible, the equation can then be solved to find part of the answer. The graphing method is commonly used for these types of systems. On a graph, these two lines can be found to intersect at one point, at two points across the parabola, or at no points.

Solving a system of one linear equation and one quadratic equation algebraically involves using the substitution method. Consider the following system: $y = x^2 + 9x + 11$; $y = 2x - 1$. Substitute the equivalent value of y from the linear equation $(2x - 1)$ into the quadratic equation. The resulting equation would be:

$$2x - 1 = x^2 + 9x + 11$$

Next, solve the resulting quadratic equation using the appropriate method—factoring, taking square roots, or using the quadratic formula. This equation can be solved by factoring:

$$0 = x^2 + 7x + 12 \rightarrow 0 = (x + 3)(x + 4) \rightarrow x + 3 = 0 \text{ or } x + 4 = 0 \rightarrow x = -3 \text{ or } x = -4$$

Next, find the corresponding y-values by substituting the x-values into the original linear equation:

$$y = 2(-4) - 1 = -9; y = 2(-3) - 1 = -7$$

Write the solutions as ordered pairs: $(-4, -9)$ and $(-3, -7)$. Finally, check the possible solutions by substituting each into both original equations. (In this case, both solutions "check out.")

## Rewriting Simple Rational Expressions

A fraction, or ratio, wherein each part is a polynomial, defines *rational expressions*. Some examples include $\frac{2x+6}{x}$, $\frac{1}{x^2-4x+8}$, and $\frac{z^2}{z+5}$. Exponents on the variables are restricted to whole numbers, which means roots and negative exponents are not included in rational expressions.

Rational expressions can be transformed by factoring. For example, the expression $\frac{x^2-5x+6}{(x-3)}$ can be rewritten by factoring the numerator to obtain $\frac{(x-3)(x-2)}{(x-3)}$. Therefore, the common binomial $(x-3)$ can cancel so that the simplified expression is:

$$\frac{(x-2)}{1} = (x-2)$$

Additionally, other rational expressions can be rewritten to take on different forms. Some may be factorable in themselves, while others can be transformed through arithmetic operations. Rational expressions are closed under addition, subtraction, multiplication, and division by a nonzero expression. *Closed* means that if any one of these operations is performed on a rational expression, the result will still be a rational expression. The set of all real numbers is another example of a set closed under all four operations.

Adding and subtracting rational expressions is based on the same concepts as adding and subtracting simple fractions. For both concepts, the denominators must be the same for the operation to take place. For example, here are two rational expressions:

$$\frac{x^3-4}{(x-3)} + \frac{x+8}{(x-3)}$$

Since the denominators are both $(x-3)$, the numerators can be combined by collecting like terms to form:

$$\frac{x^3+x+4}{(x-3)}$$

If the denominators are different, they need to be made common (the same) by using the Least Common Denominator (LCD). Each denominator needs to be factored, and the LCD contains each factor that appears in any one denominator the greatest number of times it appears in any denominator. The original expressions need to be multiplied times a form of 1, which will turn each denominator into the LCD. This process is like adding fractions with unlike denominators. It is also important when working with rational expressions to define what value of the variable makes the denominator zero. For this particular value, the expression is undefined.

Multiplication of rational expressions is performed like multiplication of fractions. The numerators are multiplied; then, the denominators are multiplied. The final fraction is then simplified. The expressions are simplified by factoring and cancelling out common terms. In the following example, the numerator of the second expression can be factored first to simplify the expression before multiplying:

$$\frac{x^2}{(x-4)} \times \frac{x^2-x-12}{2}$$

$$\frac{x^2}{(x-4)} \times \frac{(x-4)(x+3)}{2}$$

The $(x - 4)$ on the top and bottom cancel out:

$$\frac{x^2}{1} \times \frac{(x + 3)}{2}$$

Then multiplication is performed, resulting in:

$$\frac{x^3 + 3x^2}{2}$$

Dividing rational expressions is similar to the division of fractions, where division turns into multiplying by a reciprocal. So the following expression can be rewritten as a multiplication problem:

$$\frac{x^2 - 3x + 7}{x - 4} \div \frac{x^2 - 5x + 3}{x - 4}$$

$$\frac{x^2 - 3x + 7}{x - 4} \times \frac{x - 4}{x^2 - 5x + 3}$$

The $x - 4$ cancels out, leaving:

$$\frac{x^2 - 3x + 7}{x^2 - 5x + 3}$$

The final answers should always be completely simplified. If a function is composed of a rational expression, the zeros of the graph can be found from setting the polynomial in the numerator as equal to zero and solving. The values that make the denominator equal to zero will either exist on the graph as a hole or a vertical asymptote.

## Interpreting Parts of Nonlinear Expressions

When a nonlinear function is used to model a real-life scenario, some aspects of the function may be relevant, while others may not. The context of each scenario will dictate what should be used. In general, x- and y-intercepts will be points of interest. A y-intercept is the value of y when $x = 0$, and an x-intercept is the value of x when $y = 0$. Suppose a nonlinear function models the value of an investment (y) over the course of time (x). It would be relevant to determine the initial value. This initial value would be the y-intercept of the function (where time $= 0$). It would also be useful to note any point in time in which the value of the investment would be 0. These would be the x-intercepts of the function.

Another aspect of a function that is typically desired is the rate of change. This tells how fast the outputs are growing or decaying with respect to given inputs. The rate of change for a quadratic function in standard form, $y = ax^2 + bx + c$, is determined by the value of a. A positive value indicates growth, and a negative value indicates decay. The rate of change for an exponential function in standard form, $y = (a)(b^x)$, is determined by the value of b. If b is greater than 1, the function describes exponential growth, and if b is less than 1, the function describes exponential decay.

For polynomial functions, the rate of change can be estimated by the highest power of the function. Polynomial functions also include absolute and/or relative minimums and maximums. Consider functions modeling production or expenses. Maximum and minimum values would be relevant aspects of these models.

Finally, the domain and range for a function should be considered for relevance. The domain consists of all input values, and the range consists of all output values. Suppose a function models the volume of a container to be produced in relation to its height. Although the function that models the scenario may include negative values for inputs and outputs, these parts of the function would obviously not be relevant.

## Understanding the Relationship Between Zeros and Factors of Polynomials

Finding the zeros of polynomial functions is the same process as finding the solutions of polynomial equations. These are the points at which the graph of the function crosses the x-axis. As stated previously, factors can be used to find the zeros of a polynomial function. The degree of the function shows the number of possible zeros. If the highest exponent on the independent variable is 4, then the degree is 4, and the number of possible zeros is 4. If there are complex solutions, the number of roots is less than the degree.

Given the function $y = x^2 + 7x + 6$, $y$ can be set equal to zero, and the polynomial can be factored. The equation turns into $0 = (x + 1)(x + 6)$, where $x = -1$ and $x = -6$ are the zeros. Since this is a quadratic equation, the shape of the graph will be a parabola. Knowing that zeros represent the points where the parabola crosses the x-axis, the maximum or minimum point is the only other piece needed to sketch a rough graph of the function. By looking at the function in standard form, the coefficient of $x$ is positive; therefore, the parabola opens *up*. Using the zeros and the minimum, the following rough sketch of the graph can be constructed:

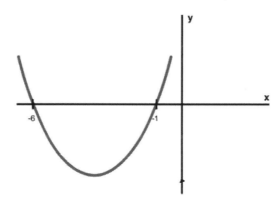

Factors for polynomials are similar to factors for integers—they are numbers, variables, or polynomials that, when multiplied together, give a product equal to the polynomial in question. One polynomial is a factor of a second polynomial if the second polynomial can be obtained from the first by multiplying by a third polynomial.

$6x^6 + 13x^4 + 6x^2$ can be obtained by multiplying together:

$$(3x^4 + 2x^2)(2x^2 + 3)$$

This means $2x^2 + 3$ and $3x^4 + 2x^2$ are factors of:

$$6x^6 + 13x^4 + 6x^2$$

In general, finding the factors of a polynomial can be tricky. However, there are a few types of polynomials that can be factored in a straightforward way.

If a certain monomial is in each term of a polynomial, it can be factored out. There are several common forms polynomials take, which if you recognize, you can solve. The first example is a perfect square trinomial. To factor this polynomial, first expand the middle term of the expression:

$$x^2 + 2xy + y^2$$

$$x^2 + xy + xy + y^2$$

Factor out a common term in each half of the expression (in this case $x$ from the left and $y$ from the right):

$$x(x + y) + y(x + y)$$

Then the same can be done again, treating $(x + y)$ as the common factor:

$$(x + y)(x + y) = (x + y)^2$$

Therefore, the formula for this polynomial is:

$$x^2 + 2xy + y^2 = (x + y)^2$$

Next is another example of a perfect square trinomial. The process is the similar, but notice the difference in sign:

$$x^2 - 2xy + y^2$$

$$x^2 - xy - xy + y^2$$

Factor out the common term on each side:

$$x(x - y) - y(x - y)$$

Factoring out the common term again:

$$(x - y)(x - y) = (x - y)^2$$

Thus:

$$x^2 - 2xy + y^2 = (x - y)^2$$

The next is known as a difference of squares. This process is effectively the reverse of binomial multiplication:

$$x^2 - y^2$$

$$x^2 - xy + xy - y^2$$

$$x(x - y) + y(x - y)$$

$$(x + y)(x - y)$$

Therefore:

$$x^2 - y^2 = (x + y)(x - y)$$

The following two polynomials are known as the sum or difference of cubes. These are special polynomials that take the form of $x^3 + y^3$ or $x^3 - y^3$. The following formula factors the sum of cubes:

$$x^3 + y^3 = (x + y)(x^2 - xy + y^2)$$

Next is the difference of cubes, but note the change in sign. The formulas for both are similar, but the order of signs for factoring the sum or difference of cubes can be remembered by using the acronym SOAP, which stands for "same, opposite, always positive." The first sign is the same as the sign in the first expression, the second is opposite, and the third is always positive. The next formula factors the difference of cubes:

$$x^3 - y^3 = (x - y)(x^2 + xy + y^2)$$

The following two examples are expansions of cubed binomials. Similarly, these polynomials always follow a pattern:

$$x^3 + 3x^2y + 3xy^2 + y^3 = (x + y)^3$$

$$x^3 - 3x^2y + 3xy^2 - y^3 = (x - y)^3$$

These rules can be used in many combinations with one another. For example, the expression $3x^3 - 24$ has a common factor of 3, which becomes:

$$3(x^3 - 8)$$

A difference of cubes still remains which can then be factored out:

$$3(x - 2)(x^2 + 2x + 4)$$

There are no other terms to be pulled out, so this expression is completely factored.

When factoring polynomials, a good strategy is to multiply the factors to check the result. Let's try another example:

$$4x^3 + 16x^2$$

Both sides of the expression can be divided by 4, and both contain $x^2$, because $4x^3$ can be thought of as $4x^2(x)$, so the common term can simply be factored out:

$$4x^2(x + 4)$$

It sometimes can be necessary to rewrite the polynomial in some clever way before applying the above rules. Consider the problem of factoring $x^4 - 1$. This does not immediately look like any of the previous polynomials. However, it's possible to think of this polynomial as $x^4 - 1 = (x^2)^2 - (1^2)^2$, and now it can be treated as a difference of squares to simplify this:

$$(x^2)^2 - (1^2)^2$$

$$(x^2)^2 - x^2 1^2 + x^2 1^2 - (1^2)^2$$

$$x^2(x^2 - 1^2) + 1^2(x^2 - 1^2)$$

$$(x^2 + 1^2)(x^2 - 1^2)$$

$$(x^2 + 1)(x^2 - 1)$$

## Understanding Nonlinear Relationships

A polynomial function consists of a monomial or sum of monomials arranged in descending exponential order. The graph of a polynomial function is a smooth continuous curve that extends infinitely on both ends.

The end behavior of the graph of a polynomial function can be determined by the degree of the function (largest exponent) and the leading coefficient (coefficient of the term with the largest exponent). If the degree is odd and the coefficient is positive, the graph falls to the left and rises to the right. If the degree is odd and the coefficient is negative, the graph rises to the left and falls to the right. If the degree is even and the coefficient is positive, the graph rises to the left and rises to the right. If the degree is even and the coefficient is negative, the graph falls to the left and falls to the right.

The y-intercept for any function is the point at which the graph crosses the y-axis. At this point, $x = 0$; therefore, to determine the y-intercept, substitute $x = 0$ into the function, and solve for y. Likewise, the x-intercepts, also called *zeros,* can be found by substituting $y = 0$ into the function and finding all solutions for x. For a given zero of a function, the graph can either pass through that point or simply touch that point (the graph turns at that zero). This is determined by the multiplicity of that zero. The multiplicity of a zero is the number of times its corresponding factor is multiplied to obtain the function in standard form. For example, $y = x^3 - 4x^2 - 3x + 18$ can be written in factored form as:

$$y = (x + 2)(x - 3)(x - 3) \text{ or } y = (x + 2)(x - 3)^2$$

The zeros of the function would be $-2$ and 3. The zero at $-2$ would have a multiplicity of 1, and the zero at 3 would have a multiplicity of 2. If a zero has an even multiplicity, then the graph touches the x-axis at that zero and turns around. If a zero has an odd multiplicity, then the graph crosses the x-axis at that zero.

The graph of a polynomial function can have several turning points (where the curve changes from rising to falling or vice versa) equal to one less than the degree of the function. For example, the function $y = 3x^5 + 2x^2 - 3x$ could have no more than four turning points.

## Using Function Notation

A function is defined as a relationship between inputs and outputs where there is only one output value for a given input. As an example, the following function is in *function notation*:

$$f(x) = 3x - 4$$

The $f(x)$ represents the output value for an input of x. If $x = 2$, the equation becomes:

$$f(2) = 3(2) - 4 = 6 - 4 = 2$$

The input of 2 yields an output of 2, forming the ordered pair $(2, 2)$. The following set of ordered pairs corresponds to the given function: $(2, 2), (0, -4), (-2, -10)$. The set of all possible inputs of a function is its *domain*, and all possible outputs is called the *range*. By definition, each member of the domain is paired with only one member of the range.

Functions can also be defined recursively. In this form, they are not defined explicitly in terms of variables. Instead, they are defined using previously-evaluated function outputs, starting with either $f(0)$ or $f(1)$. An example of a recursively-defined function is:

$$f(1) = 2, f(n) = 2f(n-1) + 2n, n > 1$$

The domain of this function is the set of all integers.

In many cases, a function can be defined by giving an equation. For instance, $f(x) = x^2$ indicates that given a value for $x$, the output of $f$ is found by squaring $x$.

Not all equations in $x$ and $y$ can be written in the form $y = f(x)$. An equation can be written in such a form if it satisfies the *vertical line test*: no vertical line meets the graph of the equation at more than a single point. In this case, $y$ is said to be a *function of x*. If a vertical line meets the graph in two places, then this equation cannot be written in the form $y = f(x)$.

The graph of a function $f(x)$ is the graph of the equation $y = f(x)$. Thus, it is the set of all pairs $(x, y)$ where $y = f(x)$. In other words, it is all pairs $(x, f(x))$. The $x$-intercepts are called the *zeros* of the function. The $y$-intercept is given by $f(0)$.

If, for a given function $f$, the only way to get $f(a) = f(b)$ is for $a = b$, then $f$ is *one-to-one*. Often, even if a function is not one-to-one on its entire domain, it is one-to-one by considering a restricted portion of the domain.

A function $f(x) = k$ for some number $k$ is called a *constant function*. The graph of a constant function is a horizontal line.

The function $f(x) = x$ is called the *identity function*. The graph of the identity function is the diagonal line pointing to the upper right at 45 degrees, $y = x$.

Given two functions, $f(x)$ *and* $g(x)$, new functions can be formed by adding, subtracting, multiplying, or dividing the functions. Any algebraic combination of the two functions can be performed, including one function being the exponent of the other. If there are expressions for $f$ and $g$, then the result can be found by performing the desired operation between the expressions. So, if $f(x) = x^2$ and $g(x) = 3x$, then:

$$f \cdot g(x) = x^2 \times 3x = 3x^3$$

Given two functions, $f(x)$ *and* $g(x)$, where the domain of $g$ contains the range of $f$, the two functions can be combined together in a process called *composition*. The function—"$g$ composed of $f$"—is written:

$$(g \circ f)(x) = g(f(x))$$

This requires the input of $x$ into $f$, then taking that result and plugging it in to the function $g$.

If $f$ is one-to-one, then there is also the option to find the function $f^{-1}(x)$, called the *inverse* of $f$. Algebraically, the inverse function can be found by writing $y$ in place of $f(x)$, and then solving for $x$. The inverse function also makes this statement true:

$$f^{-1}(f(x)) = x$$

Computing the inverse of a function $f$ entails the following procedure:

Given $f(x) = x^2$, with a domain of $x \geq 0$

$x = y^2$ is written down o find the inverse

The square root of both sides is determined to solve for $y$

Normally, this would mean $\pm\sqrt{x} = y$. However, the domain of $f$ does not include the negative numbers, so the negative option needs to be eliminated.

The result is $y = \sqrt{x}$, so $f^{-1}(x) = \sqrt{x}$, with a domain of $x \geq 0$.

A function is called *monotone* if it is either always increasing or always decreasing. For example, the functions $f(x) = 3x$ and $f(x) = -x^5$ are monotone.

An *even function* looks the same when flipped over the y-axis: $f(x) = f(-x)$. The following image shows a graphic representation of an even function.

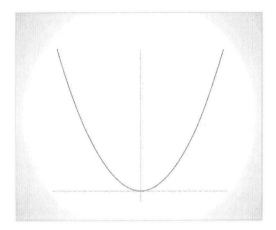

An *odd function* looks the same when flipped over the y-axis and then flipped over the x-axis: $f(x) = -f(-x)$.

The following image shows an example of an odd function.

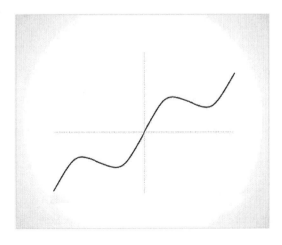

55

## Using Structure to Isolate a Quantity of Interest

Solving equations in one variable is the process of For example, in $3x - 7 = 20$, the variable $x$ needs to be isolated. Using opposite operations, the $-7$ is moved to the right side of the equation by adding seven to both sides:

$$3x - 7 + 7 = 20 + 7$$

$$3x = 27$$

Dividing by three on each side, $\frac{3x}{3} = \frac{27}{3}$, results in isolation of the variable. It is important to note that if an operation is performed on one side of the equals sign, it has to be performed on the other side to maintain equality. The solution is found to be $x = 9$. This solution can be checked for accuracy by plugging $x = 7$ in the original equation. After simplifying the equation, $20 = 20$ is found, which is a true statement.

Formulas are mathematical expressions that define the value of one quantity given the value of one or more different quantities. A formula or equation expressed in terms of one variable can be manipulated to express the relationship in terms of any other variable. The equation $y = 3x + 2$ is expressed in terms of the variable $y$. By manipulating the equation, it can be written as $x = \frac{y-2}{3}$, which is expressed in terms of the variable $x$. To manipulate an equation or formula to solve for a variable of interest, consider how the equation would be solved if all other variables were numbers. Follow the same steps for solving, leaving operations in terms of the variables instead of calculating numerical values.

The formula $P = 2l + 2w$ expresses how to calculate the perimeter of a rectangle given its length and width. To write a formula to calculate the width of a rectangle given its length and perimeter, use the previous formula relating the three variables, and solve for the variable $w$. If P and l were numerical values, this would be a two-step linear equation solved by subtraction and division. To solve the equation $P = 2l + 2w$ for $w$, first subtract $2l$ from both sides:

$$P - 2l = 2w$$

Then, divide both sides by 2:

$$\frac{P - 2l}{2} = \frac{2w}{2}$$

Or:

$$\frac{P}{2} - l = w$$

# Additional Topics in Math

## Solving Problems Using Volume Formulas

Geometry in three dimensions is similar to geometry in two dimensions. The main new feature is that three points now define a unique *plane* that passes through each of them. Three dimensional objects

can be made by putting together two dimensional figures in different surfaces. Below, some of the possible three dimensional figures will be provided, along with formulas for their volumes.

A rectangular prism is a box whose sides are all rectangles meeting at 90° angles. Such a box has three dimensions: length, width, and height. If the length is *x*, the width is *y*, and the height is *z*, then the volume is given by $V = xyz$.

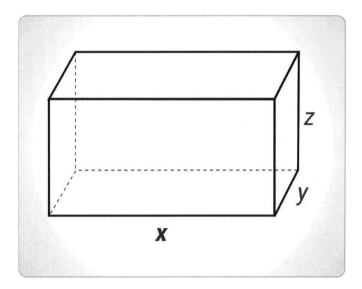

A *rectangular pyramid* is a figure with a rectangular base and four triangular sides that meet at a single vertex. If the rectangle has sides of length *x* and *y*, then the volume will be given by $V = \frac{1}{3}xyh$.

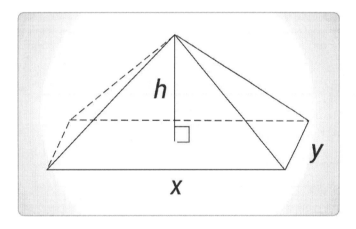

A *sphere* is a set of points all of which are equidistant from some central point. It is like a circle, but in three dimensions. The volume of a sphere of radius $r$ is given by $V = \frac{4}{3}\pi r^3$.

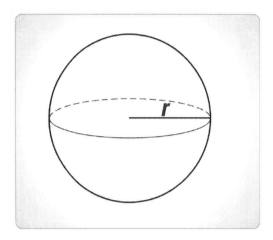

The volume of a cylinder is calculated by multiplying the area of the base (which is a circle) by the height of the cylinder. Doing so results in the equation $V = \pi r^2 h$. The volume of a cone is $^1/_3$ of the volume of a cylinder. Therefore, the formula for the volume of a cone is $\frac{1}{3}\pi r^2 h$.

## Using Trigonometric Ratios and the Pythagorean Theorem

The *Pythagorean theorem* is an important relationship between the three sides of a right triangle. It states that the square of the side opposite the right triangle, known as the *hypotenuse* (denoted as c²), is equal to the sum of the squares of the other two sides (a² + b²). Thus, a² + b² = c².

Both the trigonometric functions and the Pythagorean theorem can be used in problems that involve finding either a missing side or a missing angle of a right triangle. To do so, one must look to see what sides and angles are given and select the correct relationship that will help find the missing value. These relationships can also be used to solve application problems involving right triangles. Often, it's helpful to draw a figure to represent the problem to see what's missing.

To prove theorems about triangles, basic definitions involving triangles (e.g., equilateral, isosceles, etc.) need to be realized. Proven theorems concerning lines and angles can be applied to prove theorems about triangles. Common theorems to be proved include: the sum of all angles in a triangle equals 180 degrees; the sum of the lengths of two sides of a triangle is greater than the length of the third side; the base angles of an isosceles triangle are congruent; the line segment connecting the midpoint of two sides of a triangle is parallel to the third side and its length is half the length of the third side; and the medians of a triangle all meet at a single point.

An *isosceles triangle* contains at least two equal sides. Therefore, it must also contain two equal angles and, subsequently, contain two medians of the same length. An isosceles triangle can also be labelled as an *equilateral triangle* (which contains three equal sides and three equal angles) when it meets these conditions. In an equilateral triangle, the measure of each angle is always 60 degrees. Also within an equilateral triangle, the medians are of the same length. A *scalene triangle* can never be an equilateral or an isosceles triangle because it contains no equal sides and no equal angles. Also, medians in a scalene triangle can't have the same length. However, a *right triangle*, which is a triangle containing a

90-degree angle, can be a scalene triangle. There are two types of special right triangles. The *30-60-90 right triangle* has angle measurements of 30 degrees, 60 degrees, and 90 degrees. Because of the nature of this triangle, and through the use of the Pythagorean theorem, the side lengths have a special relationship. If $x$ is the length opposite the 30-degree angle, the length opposite the 60-degree angle is $\sqrt{3}x$, and the hypotenuse has length $2x$. The *45-45-90 right triangle* is also special as it contains two angle measurements of 45 degrees. It can be proven that, if x is the length of the two equal sides, the hypotenuse is $x\sqrt{2}$. The properties of all of these special triangles are extremely useful in determining both side lengths and angle measurements in problems where some of these quantities are given and some are not.

Trigonometric functions are also used to describe behavior in mathematics. *Trigonometry* is the relationship between the angles and sides of a triangle. *Trigonometric functions* include sine, cosine, tangent, secant, cosecant, and cotangent. The functions are defined through ratios in a right triangle. SOHCAHTOA is a common acronym used to remember these ratios, which are defined by the relationships of the sides and angles relative to the right angle. Sine is opposite over hypotenuse, cosine is adjacent over hypotenuse, and tangent is opposite over adjacent. These ratios are the reciprocals of secant, cosecant, and cotangent, respectively. Angles can be measured in degrees or radians. Here is a diagram of SOHCAHTOA:

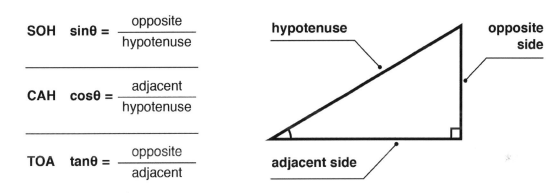

Consider the right triangle shown in this figure:

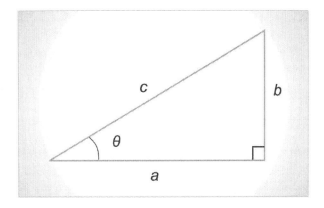

The following hold true:

- $c \sin \theta = b$

- $c \cos \theta = a$

- $\tan\theta = \dfrac{b}{a}$

- $b\csc\theta = c$

- $a\sec\theta = c$

- $\cot\theta = \dfrac{a}{b}$

A triangle that isn't a right triangle is known as an *oblique triangle*. It should be noted that even if the triangle consists of three acute angles, it is still referred to as an oblique triangle. *Oblique*, in this case, does not refer to an angle measurement. Consider the following oblique triangle:

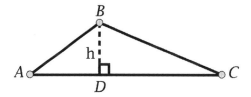

For this triangle,

$$Area = \frac{1}{2} \times base \times height = \frac{1}{2} \times AC \times BD$$

The auxiliary line drawn from the vertex B perpendicular to the opposite side AC represents the height of the triangle. This line splits the larger triangle into two smaller right triangles, which allows for the use of the trigonometric functions (specifically that $\sin A = \dfrac{h}{AB}$). Therefore,

$$Area = \frac{1}{2} \times AC \times AB \times \sin A$$

Typically the sides are labelled as the lowercase letter of the vertex that's opposite. Therefore, the formula can be written as $Area = \dfrac{1}{2}ab\sin A$. This area formula can be used to find areas of triangles when given side lengths and angle measurements, or it can be used to find side lengths or angle measurements based on a specific area and other characteristics of the triangle.

The *law of sines* and *law of cosines* are two more relationships that exist within oblique triangles. Consider a triangle with sides $a$, $b$, and $c$, and angles $A$, $B$, and $C$ opposite the corresponding sides.

The law of cosines states that:

$$c^2 = a^2 + b^2 - 2ab\cos C$$

The law of sines states that:

$$\frac{\sin A}{a} = \frac{\sin B}{b} = \frac{\sin C}{c}$$

In addition to the area formula, these two relationships can help find unknown angle and side measurements in oblique triangles.

## Complex Numbers

Complex numbers are made up of the sum of a real number and an imaginary number. Imaginary numbers are the result of taking the square root of -1, and $\sqrt{-1} = i$.

Some examples of complex numbers include $6 + 2i$, $5 - 7i$, and $-3 + 12i$. Adding and subtracting complex numbers is similar to collecting like terms. The real numbers are added together, and the imaginary numbers are added together. For example, if the problem asks to simplify the expression $6 + 2i - 3 + 7i$, the 6 and (-3) are combined to make 3, and the $2i$ and $7i$ combine to make $9i$. Multiplying and dividing complex numbers is similar to working with exponents. One rule to remember when multiplying is that $i * i = -1$. For example, if a problem asks to simplify the expression $4i(3 + 7i)$, the $4i$ should be distributed throughout the 3 and the $7i$. This leaves the final expression $12i - 28$. The 28 is negative because $i * i$ results in a negative number. The last type of operation to consider with complex numbers is the conjugate. The *conjugate* of a complex number is a technique used to change the complex number into a real number. For example, the conjugate of $4 - 3i$ is $4 + 3i$. Multiplying $(4 - 3i)(4 + 3i)$ results in $16 + 12i - 12i + 9$, which has a final answer of:

$$16 + 9 = 25$$

Complex numbers may result from solving polynomial equations using the quadratic equation. Since complex numbers result from taking the square root of a negative number, the number found under the radical in the quadratic formula—called the *determinant*—tells whether or not the answer will be real or complex. If the determinant is negative, the roots are complex. Even though the coefficients of the polynomial may be real numbers, the roots are complex.

Solving polynomials by factoring is an alternative to using the quadratic formula. For example, in order to solve $x^2 - b^2 = 0$ for $x$, it needs to be factored. It factors into:

$$(x + b)(x - b) = 0$$

The solution set can be found by setting each factor equal to zero, resulting in $x = \pm b$. When $b^2$ is negative, the factors are complex numbers. For example, $x^2 + 64 = 0$ can be factored into:

$$(x + 8i)(x - 8i) = 0$$

The two roots are then found to be $x = \pm 8i$.

When dealing with polynomials and solving polynomial equations, it is important to remember the fundamental theorem of algebra. When given a polynomial with a degree of n, the theorem states that there will be n roots. These roots may or may not be complex. For example, the following polynomial equation of degree 2 has two complex roots:

$$x^2 + 1 = 0$$

The factors of this polynomial are $(x + i)$ and $(x - i)$, resulting in the roots $x = i, -i$. As seen on the graph below, imaginary roots occur when the graph does not touch the x-axis.

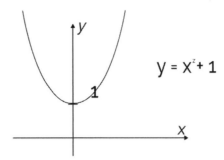

*When a graphing calculator is permitted, the graph can always confirm the number and types of roots of the polynomial.*

A polynomial identity is a true equation involving polynomials. For example:

$$x^2 - 5x + 6 = (x - 3)(x - 2)$$

This can be proved through multiplication by the FOIL method and factoring. This idea can be extended to involve complex numbers. For example:

$$i^2 = -1, x^3 + 9x = x(x^2 + 9) = x\left(x + \sqrt{3}i\right)\left(x - \sqrt{3}i\right)$$

This identity can also be proven through FOIL and factoring.

## Converting Between Degrees and Radians

A *radian* is equal to the angle that subtends the arc with the same length as the radius of the circle. It is another unit for measuring angles, in addition to degrees. The unit circle is used to describe different radian measurements and the trigonometric ratios for special angles. The circle has a center at the origin, $(0, 0)$, and a radius of 1, which can be seen below. The points where the circle crosses an axis are labeled.

The circle begins on the right-hand side of the x-axis at 0 radians. Since the circumference of a circle is $2\pi r$ and the radius $r = 1$, the circumference is $2\pi$. Zero and $2\pi$ are labeled as radian measurements at the point $(1, 0)$ on the graph. The radian measures around the rest of the circle are labeled also in relation to $\pi$; $\pi$ is at the point $(-1, 0)$, also known as 180 degrees. Since these two measurements are equal, $\pi = 180$ degrees written as a ratio can be used to convert degrees to radians or vice versa. For example, to convert 30 degrees to radians, 30 degrees $\times \frac{\pi}{180 \text{ degrees}}$ can be used to obtain $\frac{1}{6}\pi$ or $\frac{\pi}{6}$. This radian measure is a point the unit circle

The coordinates labeled on the unit circle are found based on two common right triangles. The ratios formed in the coordinates can be found using these triangles. Each of these triangles can be inserted into the circle to correspond 30, 45, and 60 degrees or $\frac{\pi}{6}, \frac{\pi}{4}$, and $\frac{\pi}{3}$ radians.

By letting the hypotenuse length of these triangles equal 1, these triangles can be placed inside the unit circle. These coordinates can be used to find the trigonometric ratio for any of the radian measurements on the circle.

Given any $(x, y)$ on the unit circle, $\sin(\theta) = y$, $\cos(\theta) = x$, and $\tan(\theta) = \frac{y}{x}$. The value $\theta$ is the angle that spans the arc around the unit circle. For example, finding $\sin\left(\frac{\pi}{4}\right)$ means finding the y-value corresponding to the angle $\theta = \frac{\pi}{4}$. The answer is $\frac{\sqrt{2}}{2}$. Finding $\cos\left(\frac{\pi}{3}\right)$ means finding the $x$-value corresponding to the angle $\theta = \frac{\pi}{3}$. The answer is $\frac{1}{2}$ or 0.5. Both angles lie in the first quadrant of the unit circle. Trigonometric ratios can also be calculated for radian measures past $\frac{\pi}{2}$, or 90 degrees. Since the same special angles can be moved around the circle, the results only differ with a change in sign. This can be seen at two points labeled in the second and third quadrant.

Trigonometric functions are periodic. Both sine and cosine have period $2\pi$. For each input angle value, the output value follows around the unit circle. Once it reaches the starting point, it continues around and around the circle. It is true that:

$$\sin(0) = \sin(2\pi) = \sin(4\pi), \text{ etc.}$$

and

$$\cos(0) = \cos(2\pi) = \cos(4\pi)$$

Tangent has period $\pi$, and its output values repeat themselves every half of the unit circle. The domain of sine and cosine are all real numbers, and the domain of tangent is all real numbers, except the points where cosine equals zero. It is also true that

$$\sin(-x) = -\sin x$$

$$\cos(-x) = \cos(x)$$

$$\tan(-x) = -\tan(x)$$

So sine and tangent are odd functions, while cosine is an even function. Sine and tangent are symmetric with respect the origin, and cosine is symmetric with respect to the y-axis.

## Applying Theorems About Circles

The *radius* of a circle is the distance from the center of the circle to any point on the circle. A *chord* of a circle is a straight line formed when its endpoints are allowed to be any two points on the circle. Many angles exist within a circle. A *central angle* is formed by using two radii as its rays and the center of the circle as its vertex. An inscribed angle is formed by using two chords as its rays, and its vertex is a point on the circle itself. Finally, a *circumscribed angle* has a vertex that is a point outside the circle and rays that intersect with the circle. Some relationships exist between these types of angles, and, in order to define these relationships, arc measure must be understood. An *arc* of a circle is a portion of the circumference. Finding the *arc measure* is the same as finding the degree measure of the central angle that intersects the circle to form the arc. The measure of an inscribed angle is half the measure of its intercepted arc. It's also true that the measure of a circumscribed angle is equal to 180 degrees minus the measure of the central angle that forms the arc in the angle.

A *tangent line* is a line that touches a curve at a single point without going through it. A *compass* and a *straight edge* are the tools necessary to construct a tangent line from a point *P* outside the circle to the circle. A tangent line is constructed by drawing a line segment from the center of the circle *O* to the point *P*, and then finding its midpoint *M* by bisecting the line segment. By using *M* as the center, a compass is used to draw a circle through points *O* and *P*. *N* is defined as the intersection of the two circles. Finally, a line segment is drawn through *P* and *N*. This is the tangent line. Each point on a circle has only one tangent line, which is perpendicular to the radius at that point. A line similar to a tangent line is a *secant line*. Instead of intersecting the circle at one point, a secant line intersects the circle at two points. A *chord* is a smaller portion of a secant line.

Here's an example:

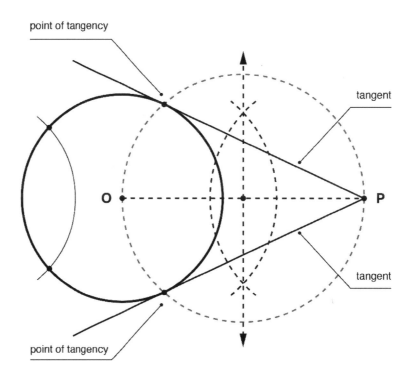

As previously mentioned, angles can be measured in radians, and 180 degrees equals π radians. Therefore, the measure of a complete circle is 2π radians. In addition to arc measure, *arc length* can also be found because the length of an arc is a portion of the circle's circumference. The following proportion is true:

$$\frac{\text{Arc measure}}{360 \text{ degrees}} = \frac{\text{arc length}}{\text{arc circumference}}$$

Arc measure is the same as the measure of the central angle, and this proportion can be rewritten as:

$$\text{arc length} = \frac{\text{central angle}}{360 \text{ degrees}} \times \text{circumference}$$

In addition, the degree measure can be replaced with radians to allow the use of both units. The arc length of a circle in radians is arc length = central angle × radius.

Note that arc length is a fractional part of circumference because $\frac{\text{central angle}}{360 \text{ degrees}} < 1$.

A *sector* of a circle is a portion of the circle that's enclosed by two radii and an arc. It resembles a piece of a pie, and the area of a sector can be derived using known definitions. The area of a circle can be

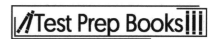

calculated using the formula $A = \pi r^2$, where $r$ is the radius of the circle. The area of a sector of a circle is a fraction of that calculation. For example, if the central angle $\theta$ is known in radians, the area of a sector is defined as:

$$A_s = \pi r^2 \frac{\theta}{2\pi} = \frac{\theta r^2}{2}$$

If the angle $\theta$ in degrees is known, the area of the sector is $A_s = \frac{\theta \pi r^2}{360}$. Finally, if the arc length $L$ is known, the area of the sector can be reduced to $A_s = \frac{rL}{2}$.

A chord is a line segment that contains endpoints on a circle. Given the radius of the circle and the central angle that inscribes the chord, the length of the chord can be determined.

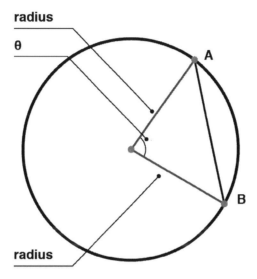

In the above figure, $\overline{AB}$ is a chord inscribed by $\angle\,\theta$. By constructing an angle bisector, the chord is also bisected at a 90 degree angle.

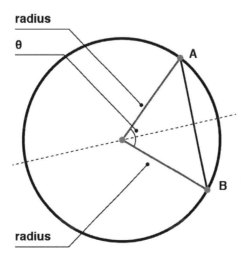

A right triangle is formed consisting of the radius as the hypotenuse, an angle equal to half the measure of $\theta$, and a side opposite to that angle with a length half the length of the chord. Given an angle and the

hypotenuse, the sin function can be used to determine the length of the side opposite the angle.
Therefore, the formula $n\frac{\theta}{2} = \frac{c}{r}$, where $c$ is the side of the triangle equal to

half of the chord, can be used. Manipulating this formula results in:

$$c = r \times \sin\frac{\theta}{2}$$

(Remember that in this formula, $c$ represents half the length of the chord, so double this length to determine the length of the chord.)

## Congruence and Similarity

Sometimes, two figures are similar, meaning they have the same basic shape and the same interior angles, but they have different dimensions. If the ratio of two corresponding sides is known, then that ratio, or scale factor, holds true for all of the dimensions of the new figure.

Here is an example of applying this principle. Suppose that Lara is 5 feet tall and is standing 30 feet from the base of a light pole, and her shadow is 6 feet long. How high is the light on the pole? To figure this, it helps to make a sketch of the situation:

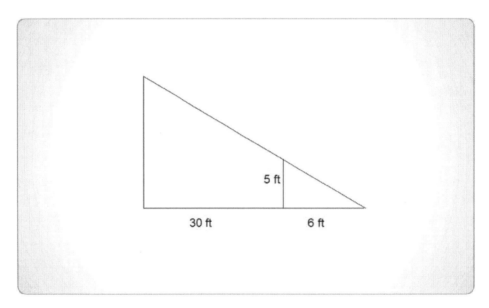

The light pole is the left side of the triangle. Lara is the 5-foot vertical line. Notice that there are two right triangles here, and that they have all the same angles as one another. Therefore, they form similar triangles. So, figure the ratio of proportionality between them.

The bases of these triangles are known. The small triangle, formed by Lara and her shadow, has a base of 6 feet. The large triangle, formed by the light pole along with the line from the base of the pole out to the end of Lara's shadow is $30 + 6 = 36$ feet long. So, the ratio of the big triangle to the little triangle will be $\frac{36}{6} = 6$. The height of the little triangle is 5 feet. Therefore, the height of the big triangle will be $6 \times 5 = 30$ feet, meaning that the light is 30 feet up the pole.

Notice that the perimeter of a figure changes by the ratio of proportionality between two similar figures, but the area changes by the *square* of the ratio. This is because if the length of one side is doubled, the area is quadrupled.

As an example, suppose two rectangles are similar, but the edges of the second rectangle are three times longer than the edges of the first rectangle. The area of the first rectangle is 10 square inches. How much more area does the second rectangle have than the first?

To answer this, note that the area of the second rectangle is $3^2 = 9$ times the area of the first rectangle, which is 10 square inches. Therefore, the area of the second rectangle is going to be $9 \times 10 = 90$ square inches. This means it has $90 - 10 = 80$ square inches more area than the first rectangle.

As a second example, suppose $X$ and $Y$ are similar right triangles. The hypotenuse of $X$ is 4 inches. The area of $Y$ is $\frac{1}{4}$ the area of $X$. What is the hypotenuse of $Y$?

First, realize the area has changed by a factor of $\frac{1}{4}$. The area changes by a factor that is the *square* of the ratio of changes in lengths, so the ratio of the lengths is the square root of the ratio of areas. That means that the ratio of lengths must be is $\sqrt{\frac{1}{4}} = \frac{1}{2}$, and the hypotenuse of $Y$ must be $\frac{1}{2} \times 4 = 2$ inches.

Volumes between similar solids change like the cube of the change in the lengths of their edges. Likewise, if the ratio of the volumes between similar solids is known, the ratio between their lengths is known by finding the cube root of the ratio of their volumes.

For example, suppose there are two similar rectangular pyramids $X$ and $Y$. The base of $X$ is 1 inch by 2 inches, and the volume of $X$ is 8 inches. The volume of $Y$ is 64 inches. What are the dimensions of the base of $Y$?

To answer this, first find the ratio of the volume of $Y$ to the volume of $X$. This will be given by $\frac{64}{8} = 8$. Now the ratio of lengths is the cube root of the ratio of volumes, or $\sqrt[3]{8} = 2$. So, the dimensions of the base of $Y$ must be 2 inches by 4 inches.

A *rigid motion* is a transformation that preserves distance and length. Every line segment in the resulting image is congruent to the corresponding line segment in the pre-image. Congruence between two figures means a series of transformations (or a rigid motion) can be defined that maps one of the figures onto the other. Basically, two figures are congruent if they have the same shape and size.

A shape is dilated, or a *dilation* occurs, when each side of the original image is multiplied by a given scale factor. If the scale factor is less than 1 and greater than 0, the dilation contracts the shape, and the resulting shape is smaller. If the scale factor equals 1, the resulting shape is the same size, and the dilation is a rigid motion. Finally, if the scale factor is greater than 1, the resulting shape is larger and the dilation expands the shape. The *center of dilation* is the point where the distance from it to any point on the new shape equals the scale factor times the distance from the center to the corresponding point in the pre-image. Dilation isn't an isometric transformation because distance isn't preserved. However, angle measure, parallel lines, and points on a line all remain unchanged.

Two figures are congruent if there is a rigid motion that can map one figure onto the other. Therefore, all pairs of sides and angles within the image and pre-image must be congruent. For example, in triangles, each pair of the three sides and three angles must be congruent. Similarly, in two four-sided figures, each pair of the four sides and four angles must be congruent.

Two figures are *similar* if there is a combination of translations, reflections, rotations, and dilations, which maps one figure onto the other. The difference between congruence and similarity is that dilation can be used in similarity. Therefore, side lengths between each shape can differ. However, angle measure must be preserved within this definition. If two polygons differ in size so that the lengths of corresponding line segments differ by the same factor, but corresponding angles have the same measurement, they are similar.

There are five theorems to show that triangles are congruent when it's unknown whether each pair of angles and sides are congruent. Each theorem is a shortcut that involves different combinations of sides and angles that must be true for the two triangles to be congruent. For example, *side-side-side (SSS)* states that if all sides are equal, the triangles are congruent. *Side-angle-side (SAS)* states that if two pairs of sides are equal and the included angles are congruent, then the triangles are congruent. Similarly, *angle-side-angle (ASA)* states that if two pairs of angles are congruent and the included side lengths are equal, the triangles are similar. *Angle-angle-side (AAS)* states that two triangles are congruent if they have two pairs of congruent angles and a pair of corresponding equal side lengths that aren't included. Finally, *hypotenuse-leg (HL)* states that if two right triangles have equal hypotenuses and an equal pair of shorter sides, then the triangles are congruent. An important item to note is that angle-angle-angle *(AAA)* is not enough information to have congruence. It's important to understand why these rules work by using rigid motions to show congruence between the triangles with the given properties. For example, three reflections are needed to show why *SAS* follows from the definition of congruence.

If two angles of one triangle are congruent with two angles of a second triangle, the triangles are similar. This is because, within any triangle, the sum of the angle measurements is 180 degrees. Therefore, if two are congruent, the third angle must also be congruent because their measurements are equal. Three congruent pairs of angles mean that the triangles are similar.

The criteria needed to prove triangles are congruent involves both angle and side congruence. Both pairs of related angles and sides need to be of the same measurement to use congruence in a proof. The criteria to prove similarity in triangles involves proportionality of side lengths. Angles must be congruent in similar triangles; however, corresponding side lengths only need to be a constant multiple of each other. Once similarity is established, it can be used in proofs as well. Relationships in geometric figures other than triangles can be proven using triangle congruence and similarity. If a similar or congruent triangle can be found within another type of geometric figure, their criteria can be used to prove a relationship about a given formula. For instance, a rectangle can be broken up into two congruent triangles.

## Similarity, Triangles, and Trigonometric Ratios

Within similar triangles, corresponding sides are proportional, and angles are congruent. In addition, within similar triangles, the ratio of the side lengths is the same. This property is true even if side lengths are different. Within right triangles, trigonometric ratios can be defined for the acute angle within the triangle. The functions are defined through ratios in a right triangle. Sine of acute angle, A, is opposite over hypotenuse, cosine is adjacent over hypotenuse, and tangent is opposite over adjacent. Note that expanding or shrinking the triangle won't change the ratios. However, changing the angle measurements will alter the calculations.

Angles that add up to 90 degrees are *complementary*. Within a right triangle, two complementary angles exist because the third angle is always 90 degrees. In this scenario, the *sine* of one of the complementary angles is equal to the *cosine* of the other angle. The opposite is also true. This relationship exists because sine and cosine will be calculated as the ratios of the same side lengths.

Two lines can be parallel, perpendicular, or neither. If two lines are parallel, they have the same slope. This is proven using the idea of similar triangles. Consider the following diagram with two parallel lines, L1 and L2:

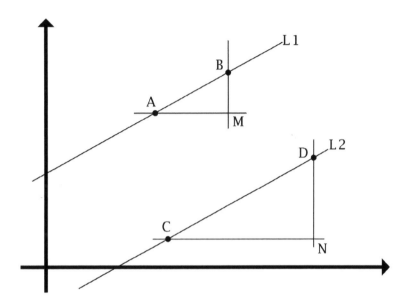

A and B are points on L1, and C and D are points on L2. Right triangles are formed with vertex M and N where lines BM and DN are parallel to the *y*-axis and AM and CN are parallel to the *x*-axis. Because all three sets of lines are parallel, the triangles are similar. Therefore, $\frac{BM}{DN} = \frac{MA}{NC}$. This shows that the rise/run is equal for lines L1 and L2. Hence, their slopes are equal.

Another similar theorem states that if there is a line parallel to one side of a triangle, and it intersects the other sides of the triangle, then the sides that are intersected are divided proportionally.

Secondly, if two lines are perpendicular, the product of their slopes equals -1. This means that their slopes are negative reciprocals of each other. Consider two perpendicular lines, *l* and *n*:

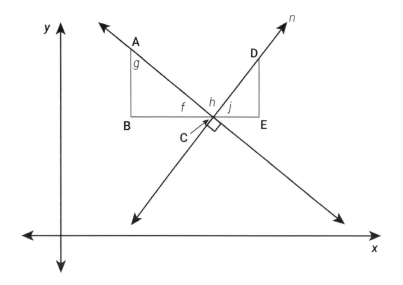

Right triangles ABC and CDE are formed so that lines BC and CE are parallel to the *x*-axis, and AB and DE are parallel to the *y*-axis. Because line BE is a straight line, angles:

$$f + h + i = 180 \text{ degrees}$$

However, angle *h* is a right angle, so:

$$f + j = 90 \text{ degrees}$$

By construction, $f + g = 90$, which means that $g = j$. Therefore, because angles $B = E$ and $g = j$, the triangles are similar and $\frac{AB}{BC} = \frac{CE}{DE}$. Because slope is equal to rise/run, the slope of line *l* is $-\frac{AB}{BC}$ and the slope of line *n* is $\frac{DE}{CE}$. Multiplying the slopes together gives:

$$-\frac{AB}{BC} \times \frac{DE}{CE} = -\frac{CE}{DE} \times \frac{DE}{CE} = -1$$

This proves that the product of the slopes of two perpendicular lines equals -1. Both parallel and perpendicular lines can be integral in many geometric proofs, so knowing and understanding their properties is crucial for problem-solving.

## Creating Equations to Solve Problems Involving Circles

A *circle* can be defined as the set of all points that are the same distance (known as the radius, *r*) from a single point C (known as the center of the circle). The center has coordinates $(h, k)$, and any point on the circle can be labelled with coordinates $(x, y)$.

As shown below, a *right triangle* is formed with these two points:

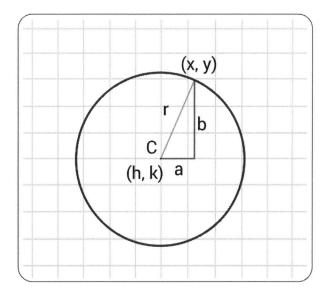

The Pythagorean theorem states that:

$$a^2 + b^2 = r^2$$

However, $a$ can be replaced by $|x - h|$ and $b$ can be replaced by $|y - k|$ by using the *distance formula* which is:

$$d = \sqrt{(x_2 - x_1)^2 + (y_2 - y_1)^2}$$

That substitution results in:

$$(x - h)^2 + (y - k)^2 = r^2$$

This is the formula for finding the equation of any circle with a center $(h, k)$ and a radius $r$. Note that sometimes $c$ is used instead of $r$.

Circles aren't always given in the form of the circle equation where the center and radius can be seen so easily. Oftentimes, they're given in the more general format of:

$$ax^2 + by^2 + cx + dy + e = 0$$

This can be converted to the center-radius form using the algebra technique of completing the square in both variables. First, the constant term is moved over to the other side of the equals sign, and then the $x$ and $y$ variable terms are grouped together. Then the equation is divided through by $a$ and, because this is the equation of a circle, $a = b$. At this point, the $x$-term coefficient is divided by 2, squared, and then added to both sides of the equation. This value is grouped with the $x$ terms. The same steps then need to be completed with the $y$-term coefficient. The trinomial in both $x$ and $y$ can now be factored into a square of a binomial, which gives both $(x - h)^2$ and $(y - k)^2$.

# SAT Math Practice Test #1

1. Which of the following inequalities is equivalent to $3 - \frac{1}{2}x \geq 2$?

    a. $x \geq 2$

    b. $x \leq 2$

    c. $x \geq 1$

    d. $x \leq 1$

2. If $g(x) = x^3 - 3x^2 - 2x + 6$ and $f(x) = 2$, then what is $g(f(x))$?

    a. -26

    b. 6

    c. $2x^3 - 6x^2 - 4x + 12$

    d. -2

3. What is the definition of a factor of the number 36?

    a. A number that can be divided by 36 and have no remainder

    b. A number that 36 can be divided by and have no remainder

    c. A prime number that is multiplied times 36

    d. An even number that is multiplied times 36

4. What are the coordinates of the focus of the parabola $y = -9x^2$?

    a. $(-3, 0)$

    b. $\left(-\frac{1}{36}, 0\right)$

    c. $(0, -3)$

    d. $\left(0, -\frac{1}{36}\right)$

5. What is the volume of a cube with the side equal to 3 inches?

    a. 6 in$^3$

    b. 27 in$^3$

    c. 9 in$^3$

    d. 3 in$^3$

6. What is the volume of a rectangular prism with the height of 3 centimeters, a width of 5 centimeters, and a depth of 11 centimeters?

    a. 19 cm$^3$

    b. 165 cm$^3$

    c. 225 cm$^3$

    d. 150 cm$^3$

7. What is the volume of a cylinder, in terms of $\pi$, with a radius of 5 inches and a height of 10 inches?

    a. 250 $\pi$ in$^3$

    b. 50 $\pi$ in$^3$

    c. 100 $\pi$ in$^3$

    d. 200 $\pi$ in$^3$

8. What is the solution to the following system of equations?

$$x^2 - 2x + y = 8$$
$$x - y = -2$$

a. $(-2, 3)$
b. There is no solution.
c. $(-2, 0) \ (1, 3)$
d. $(-2, 0) \ (3, 5)$

9. An equation for the line passing through the origin and the point $(2, 1)$ is
a. $y = 2x$
b. $y = \frac{1}{2}x$
c. $y = x - 2$
d. $2y = x + 1$

10. A rectangle was formed out of pipe cleaner. Its length was $\frac{1}{2}$ of a foot and its width was $\frac{11}{2}$ inches. What is its area in square inches?
a. $\frac{11}{4}$ in$^2$
b. $\frac{11}{2}$ in$^2$
c. 22 in$^2$
d. 33 in$^2$

11. What type of function is modeled by the values in the following table?

| $x$ | $f(x)$ |
|---|---|
| 1 | 2 |
| 2 | 4 |
| 3 | 8 |
| 4 | 16 |
| 5 | 32 |

a. Linear
b. Exponential
c. Quadratic
d. Cubic

12. Two cards are drawn from a shuffled deck of 52 cards. What's the probability that both cards are Kings if the first card isn't replaced after it's drawn?
a. $\frac{1}{169}$
b. $\frac{1}{221}$
c. $\frac{1}{13}$
d. $\frac{4}{13}$

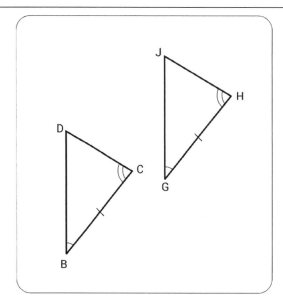

13. In the image above, what is demonstrated by the two triangles?
    a. According to Side-Side-Side, the triangles are congruent.
    b. According to Angle-Angle-Angle, the triangles are congruent.
    c. According to Angle-Side-Angle, the triangles are congruent.
    d. There is not enough information to prove the two triangles are congruent.

14. Which expression is the same as six less than three times the sum of twice a number and one?
    a. $2x + 1 - 6$
    b. $3x + 1 - 6$
    c. $3(x + 1) - 6$
    d. $3(2x + 1) - 6$

15. If $\sqrt{5 + x} = 5$, what is $x$?
    a. 10
    b. 15
    c. 20
    d. 25

16. Which of the following polynomials is equal to $(2x - 4y)^2$?
    a. $4x^2 - 16xy + 16y^2$
    b. $4x^2 - 8xy + 16y^2$
    c. $4x^2 - 16xy - 16y^2$
    d. $2x^2 - 8xy + 8y^2$

17. What are the zeros of $f(x) = x^2 + 4$?
    a. $x = -4$
    b. $x = \pm 2i$
    c. $x = \pm 2$
    d. $x = \pm 4i$

18. Which of the following shows the correct result of simplifying the following expression:

$$(7n + 3n^3 + 3) + (8n + 5n^3 + 2n^4)$$

a. $9n^4 + 15n - 2$
b. $2n^4 + 5n^3 + 15n - 2$
c. $9n^4 + 8n^3 + 15n$
d. $2n^4 + 8n^3 + 15n + 3$

19. What is the simplified result of $\frac{15}{23} \times \frac{54}{127}$?

a. $\frac{810}{2,921}$

b. $\frac{81}{292}$

c. $\frac{69}{150}$

d. $\frac{810}{2929}$

20. What is the product of the following expression?

$$(4x - 8)(5x^2 + x + 6)$$

a. $20x^3 - 36x^2 + 16x - 48$
b. $6x^3 - 41x^2 + 12x + 15$
c. $20x^3 + 11x^2 - 37x - 12$
d. $2x^3 - 11x^2 - 32x + 20$

21. What is the solution for the following equation?

$$\frac{x^2 + x - 30}{x - 5} = 11$$

a. $x = -6$
b. There is no solution.
c. $x = 16$
d. $x = 5$

22. If $x$ is not zero, then $\frac{3}{x} + \frac{5u}{2x} - \frac{u}{4} =$

a. $\frac{12 + 10u - ux}{4x}$

b. $\frac{3 + 5u - ux}{x}$

c. $\frac{12x + 10u + ux}{4x}$

d. $\frac{12 + 10u - u}{4x}$

23. What are the zeros of the function: $f(x) = x^3 + 4x^2 + 4x$?
    a. -2
    b. 0, -2
    c. 2
    d. 0, 2

24. Is the following function even, odd, neither, or both?

$$y = \frac{1}{2}x^4 + 2x^2 - 6$$

    a. Even
    b. Odd
    c. Neither
    d. Both

25. Which of the following formulas would correctly calculate the perimeter of a legal-sized piece of paper that is 14 inches long and $8\frac{1}{2}$ inches wide?
    a. $P = 14 + 8\frac{1}{2}$

    b. $P = 14 + 8\frac{1}{2} + 14 + 8\frac{1}{2}$

    c. $P = 14 \times 8\frac{1}{2}$

    d. $P = 14 \times \frac{17}{2}$

26. A grocery store is selling individual bottles of water, and each bottle contains 750 milliliters of water. If 12 bottles are purchased, what conversion will correctly determine how many liters that customer will take home?
    a. 100 milliliters equals 1 liter
    b. 1,000 milliliters equals 1 liter
    c. 1,000 liters equals 1 milliliter
    d. 10 liters equals 1 milliliter

27. Given the following triangle, what's the length of the missing side? Round the answer to the nearest tenth.

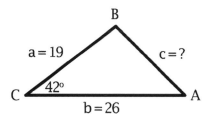

    a. 17.0
    b. 17.4
    c. 18.0
    d. 18.4

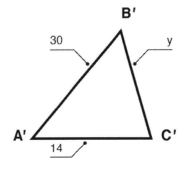
28. For the following similar triangles, what are the values of $x$ and $y$ (rounded to one decimal place)?

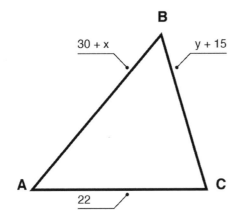

a. $x = 16.5, y = 25.1$
b. $x = 19.5, y = 24.1$
c. $x = 17.1, y = 26.3$
d. $x = 26.3, y = 17.1$

29. What are the center and radius of a circle with equation $4x^2 + 4y^2 - 16x - 24y + 51 = 0$?
a. Center $(3, 2)$ and radius $1/2$
b. Center $(2, 3)$ and radius $1/2$
c. Center $(3, 2)$ and radius $1/4$
d. Center $(2, 3)$ and radius $1/4$

30. What is the solution to $(2 \times 20) \div (7 + 1) + (6 \times 0.01) + (4 \times 0.001)$?
a. 5.064
b. 5.64
c. 5.0064
d. 48.064

31. A piggy bank contains 12 dollars' worth of nickels. A nickel weighs 5 grams, and the empty piggy bank weighs 1050 grams. What is the total weight of the full piggy bank?
a. 1,110 grams
b. 1,200 grams
c. 2,250 grams
d. 2,200 grams

32. Last year, the New York City area received approximately $27\frac{3}{4}$ inches of snow. The Denver area received approximately 3 times as much snow as New York City. How much snow fell in Denver?
a. 60 inches
b. $27\frac{1}{4}$ inches
c. $9\frac{1}{4}$ inches
d. $83\frac{1}{4}$ inches

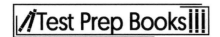
33. If $-3(x + 4) \geq x + 8$, what is the value of $x$?
   a. $x = 4$
   b. $x \geq 2$
   c. $x \geq -5$
   d. $x \leq -5$

34. Karen gets paid a weekly salary and a commission for every sale that she makes. The table below shows the number of sales and her pay for different weeks.

| Sales | 2 | 7 | 4 | 8 |
|-------|-----|-----|-----|-----|
| Pay | $380 | $580 | $460 | $620 |

Which of the following equations represents Karen's weekly pay?
   a. $y = 90x + 200$
   b. $y = 90x - 200$
   c. $y = 40x + 300$
   d. $y = 40x - 300$

35. The square and circle have the same center. The circle has a radius of $r$. What is the area of the shaded region?

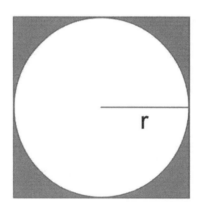

   a. $r^2 - \pi r^2$
   b. $4r^2 - 2\pi r$
   c. $(4 - \pi)r^2$
   d. $(\pi - 1)r^2$

36. The graph shows the position of a car over a 10-second time interval. Which of the following is the correct interpretation of the graph for the interval 1 to 3 seconds?

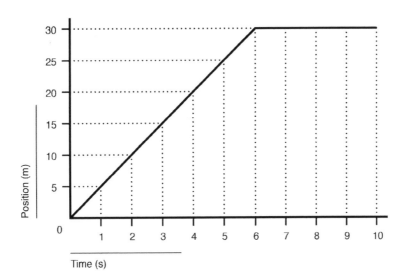

Time (s)

a. The car remains in the same position.
b. The car is traveling at a speed of 5 m/s.
c. The car is traveling up a hill.
d. The car is traveling at 5 mph.

37. Which of the ordered pairs below is a solution to the following system of inequalities?

$$y > 2x - 3$$
$$y < -4x + 8$$

a. (4, 5)
b. (-3, -2)
c. (3, -1)
d. (5, 2)

38. Which equation best represents the scatterplot below?

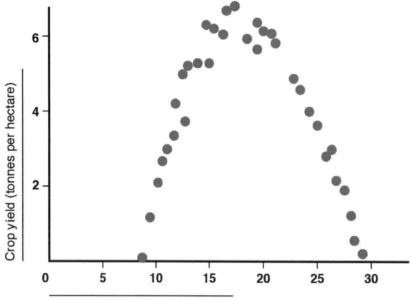

a. $y = 3x - 4$
b. $y = 2x^2 + 7x - 9$
c. $y = (3)(4^x)$
d. $y = -\frac{1}{14}x^2 + 2x - 8$

39. What is the solution to $9 \times 9 \div 9 + 9 - 9 \div 9$?
   a. 0
   b. 17
   c. 81
   d. 9

40. What is the solution to the radical equation $\sqrt[3]{2x + 11} + 9 = 12$?
   a. -8
   b. 8
   c. 0
   d. 12

41. The hospital has a nurse to patient ratio of 1:25. If there is a maximum of 325 patients admitted at a time, how many nurses are there?
   a. 13 nurses
   b. 25 nurses
   c. 325 nurses
   d. 12 nurses

42. A hospital has a bed to room ratio of 2 to 1. If there are 145 rooms, how many beds are there?
    a. 145 beds
    b. 2 beds
    c. 90 beds
    d. 290 beds

43. If $\frac{2x}{5} - 1 = 59$, what is the value of $x$?
    a. 60
    b. 145
    c. 150
    d. 115

44. A National Hockey League store in the state of Michigan advertises 50% off all items. Sales tax in Michigan is 6%. How much would a hat originally priced at $32.99 and a jersey originally priced at $64.99 cost during this sale? Round to the nearest penny.
    a. $97.98
    b. $103.86
    c. $51.93
    d. $48.99

45. Store brand coffee beans cost $1.23 per pound. A local coffee bean roaster charges $1.98 per 1 and a half pounds. How much more would 5 pounds from the local roaster cost than 5 pounds of the store brand?
    a. $0.55
    b. $1.55
    c. $1.45
    d. $0.45

46. Paint Inc. charges $2000 for painting the first 1,800 feet of trim on a house and $1.00 per foot for each foot after. How much would it cost to paint a house with 3125 feet of trim?
    a. $3125
    b. $2000
    c. $5125
    d. $3325

## No Calculator Questions

47. A bucket can hold 11.4 liters of water. A kiddie pool needs 35 gallons of water to be full. How many times will the bucket need to be filled to fill the kiddie pool?
    a. 12
    b. 35
    c. 11
    d. 45

48. In Jim's school, there are 3 girls for every 2 boys. There are 650 students in total. Using this information, how many students are girls?
    a. 260
    b. 130
    c. 65
    d. 390

49. What is the volume of a pyramid, with a square base whose side is 6 inches, and the height is 9 inches?
    a. 324 in$^3$
    b. 72 in$^3$
    c. 108 in$^3$
    d. 18 in$^3$

50. Convert $\frac{2}{9}$ to a percentage.
    a. 22%
    b. 4.5%
    c. 450%
    d. 0.22%

51. What is the volume of a cone, in terms of $\pi$, with a radius of 10 centimeters and height of 12 centimeters?
    a. 400 cm$^3$
    b. 200 cm$^3$
    c. 120 cm$^3$
    d. 140 cm$^3$

52. What is 3 out of 8 expressed as a percent?
    a. 37.5%
    b. 37%
    c. 26.7%
    d. 2.67%

53. The area of a given rectangle is 24 square centimeters. If the measure of each side is multiplied by 3, what is the area of the new figure?
    a. 48 cm$^2$
    b. 72 cm$^2$
    c. 216 cm$^2$
    d. 13,824 cm$^2$

54. If $4x - 3 = 5$, what is the value of $x$?

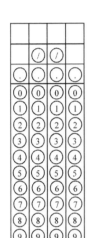

55. What is the solution to $4 \times 7 + (25 - 21)^2 \div 2$?

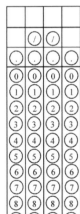

56. What is the solution to the following expression?

$$\left(\sqrt{36} \times \sqrt{16}\right) - 3^2$$

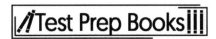 

57. What is the overall median of Dwayne's current test scores: 78, 92, 83, 97?

58. The total perimeter of a rectangle is 36 cm. If the length is 12 cm, what is the width?

# Answer Explanations #1

**1. B:** To simplify this inequality, subtract 3 from both sides to get $-\frac{1}{2}x \geq -1$. Then, multiply both sides by -2 (remembering this flips the direction of the inequality) to get $x \leq 2$.

**2. D:** This problem involves a composition function, where one function is plugged into the other function. In this case, the $f(x)$ function is plugged into the $g(x)$ function for each $x$-value. The composition equation becomes:

$$g(f(x)) = 2^3 - 3(2)^2 - 2(2) + 6$$

Simplifying the equation gives the answer:

$$g(f(x)) = 8 - 3(4) - 2(2) + 6$$

$$8 - 12 - 4 + 6 = -2$$

**3. B:** A factor of 36 is any number that can be divided into 36 and have no remainder.

$$36 = 36 \times 1, 18 \times 2, 9 \times 4, \text{and } 6 \times 6$$

Therefore, it has 7 unique factors: 36, 18, 9, 6, 4, 2, and 1.

**4. D:** A parabola of the form $y = \frac{1}{4f}x^2$ has a focus $(0, f)$.

Because $y = -9x^2$, set $-9 = \frac{1}{4f}$.

Solving this equation for $f$ results in $f = -\frac{1}{36}$. Therefore, the coordinates of the focus are $\left(0, -\frac{1}{36}\right)$.

**5. B:** The volume of a cube is the length of the side cubed, and 3 inches cubed is 27 in$^3$.

Choice $A$ is not the correct answer because that is $2 \times 3$ inches.

Choice $C$ is not the correct answer because that is $3 \times 3$ inches, and Choice $D$ is not the correct answer because there was no operation performed.

**6. B:** The volume of a rectangular prism is the $length \times width \times height$, and $3\ cm \times 5\ cm \times 11\ cm$ is 165 cm$^3$.

Choice $A$ is not the correct answer because that is $3\ cm + 5\ cm + 11\ cm$. Choice $C$ is not the correct answer because that is $15^2$. Choice $D$ is not the correct answer because that is $3cm \times 5cm \times 10cm$.

**7. A:** The volume of a cylinder is $\pi r^2 h$, and $\pi \times 5^2 \times 10$ is $250\ \pi\ in^3$.

Choice $B$ is not the correct answer because that is $5^2 \times 2\pi$. Choice $C$ is not the correct answer since that is $5in \times 10\pi$. Choice $D$ is not the correct answer because that is $10^2 \times 2in$.

**8. D:** This system of equations involves one quadratic function and one linear function, as seen from the degree of each equation. One way to solve this is through substitution. Solving for y in the second equation yields $y = x + 2$.

Plugging this equation in for the y of the quadratic equation yields:

$$x^2 - 2x + x + 2 = 8$$

Simplifying the equation, it becomes $x^2 - x + 2 = 8$.

Setting this equal to zero and factoring, it becomes:

$$x^2 - x - 6 = 0 = (x - 3)(x + 2)$$

Solving these two factors for x gives the zeros $x = 3, -2$. To find the y-value for the point, each number can be plugged in to either original equation. Solving each one for y yields the points $(3, 5)$ and $(-2, 0)$.

**9. B:** The slope will be given by $\frac{1-0}{2-0} = \frac{1}{2}$.

The y-intercept will be 0, since it passes through the origin. Using slope-intercept form, the equation for this line is $y = \frac{1}{2}x$.

**10. D:** Recall the formula for area, area = length × width. The answer must be in square inches, so all values must be converted to inches. Half of a foot is equal to 6 inches. Therefore, the area of the rectangle is equal to:

$$6 \text{ in} \times \frac{11}{2} \text{ in} = \frac{66}{2} \text{ in}^2 = 33 \text{ in}^2$$

**11. B:** The table shows values that are increasing exponentially. The differences between the inputs are the same, while the differences in the outputs are changing by a factor of 2. The values in the table can be modeled by the equation $f(x) = 2^x$.

**12. B:** For the first card drawn, the probability of a King being pulled is $\frac{4}{52}$. Since this card isn't replaced, if a King is drawn first the probability of a King being drawn second is $\frac{3}{51}$. The probability of a King being drawn in both the first and second draw is the product of the two probabilities:

$$\frac{4}{52} \times \frac{3}{51} = \frac{12}{2652}$$

This fraction, when divided by 12, equals $\frac{1}{221}$.

**13. C:** The picture demonstrates Angle-Side-Angle congruence. Choice A and B are incorrect because the picture does not show Side-Side-Side congruence and angles alone cannot prove congruence. Choice D is not the correct answer because there is already enough information to prove congruence.

**14. D:** The expression is three times the sum of twice a number and 1, which is $3(2x + 1)$. Then, 6 is subtracted from this expression.

**15. C:** To solve this equation, square both sides to eliminate the radical, resulting in $x + 5 = 25$. Subtracting 5 from both sides to solve for $x$ gives $x = 20$.

**16. A:** To expand a squared binomial, it's necessary to use the FOIL (First, Outer, Inner, Last) method:

$$(2x - 4y)^2$$

$$2x \times 2x + 2x(-4y) + (-4y)(2x) + (-4y)(-4y)$$

$$4x^2 - 8xy - 8xy + 16y^2$$

$$4x^2 - 16xy + 16y^2$$

**17. B:** The zeros of this function can be found by using the quadratic formula:

$$x = \frac{-b \pm \sqrt{b^2 - 4ac}}{2a}$$

Identifying $a$, $b$, and $c$ can also be done from the equation because it is in standard form. The formula becomes:

$$x = \frac{0 \pm \sqrt{0^2 - 4(1)(4)}}{2(1)} = \frac{\sqrt{-16}}{2}$$

Since there is a negative underneath the radical, the answer is a complex number:

$$x = \pm 2i$$

**18. D:** The expression is simplified by collecting like terms. Terms with the same variable and exponent are like terms, and their coefficients can be added.

**19. A:** To simplify this expression, first line up the fractions:

$$\frac{15}{23} \times \frac{54}{127}$$

Multiply across the top and across the bottom to find the numerator and denominator:

$$\frac{15 \times 54}{23 \times 127} = \frac{810}{2921}$$

Because the numerator and denominator do not share any common factors, the resulting fraction cannot be reduced.

**20. A:** Finding the product means distributing one polynomial to the other so that each term in the first is multiplied by each term in the second. Then, like terms can be collected. Multiplying the factors yields the expression:

$$20x^3 + 4x^2 + 24x - 40x^2 - 8x - 48$$

Collecting like terms means adding the $x^2$ terms and adding the $x$ terms. The final answer after simplifying the expression is:

$$20x^3 - 36x^2 + 16x - 48$$

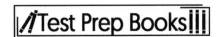 

**21. B:** The equation can be solved by factoring the numerator into $(x + 6)(x - 5)$. Since that same factor $(x - 5)$ exists on top and bottom, that factor cancels. This leaves the equation $x + 6 = 11$. Solving the equation gives the answer $x = 5$. When this value is plugged into the equation, it yields a zero in the denominator of the fraction. Since this is undefined, there is no solution.

**22. A:** The common denominator here will be $4x$. Rewrite these fractions as:

$$\frac{3}{x} + \frac{5u}{2x} - \frac{u}{4}$$

$$\frac{12}{4x} + \frac{10u}{4x} - \frac{ux}{4x}$$

$$\frac{12 + 10u - ux}{4x}$$

**23. B:** There are two zeros for the given function. They are $x = 0, -2$. The zeros can be found a number of ways, but this particular equation can be factored into:

$$f(x) = x(x^2 + 4x + 4) = x(x + 2)(x + 2)$$

By setting each factor equal to zero and solving for $x$, there are two solutions. On a graph, these zeros can be seen where the line crosses the $x$-axis.

**24. A:** The equation is *even* because $f(-x) = f(x)$. Plugging in a negative value will result in the same answer as when plugging in the positive of that same value. The function:

$$f(-2) = \frac{1}{2}(-2)^4 + 2(-2)^2 - 6$$

$$8 + 8 - 6 = 10$$

This function yields the same value as:

$$f(2) = \frac{1}{2}(2)^4 + 2(2)^2 - 6$$

$$8 + 8 - 6 = 10$$

**25. B:** The perimeter of a rectangle is the sum of all four sides. Therefore, the answer is:

$$P = 14 + 8\frac{1}{2} + 14 + 8\frac{1}{2}$$

$$14 + 14 + 8 + \frac{1}{2} + 8 + \frac{1}{2}$$

45 square inches

**26. B:** $12 \times 750 = 9{,}000$. Therefore, there are 9,000 milliliters of water, which must be converted to liters. 1,000 milliliters equals 1 liter; therefore, 9 liters of water are purchased.

**27. B:** Because this isn't a right triangle, SOHCAHTOA can't be used. However, the law of cosines can be used. Therefore:

$$c^2 = a^2 + b^2 - 2ab \cos C$$

$$19^2 + 26^2 - 2 \times 19 \times 26 \times \cos 42°$$

$$302.773$$

Taking the square root and rounding to the nearest tenth results in $c = 17.4$.

**28. C:** Because the triangles are similar, the lengths of the corresponding sides are proportional. Therefore:

$$\frac{30 + x}{30} = \frac{22}{14} = \frac{y + 15}{y}$$

This results in the equation $14(30 + x) = 22 \times 30$ which, when solved, gives $x = 17.1$. The proportion also results in the equation $14(y + 15) = 22y$ which, when solved, gives $y = 26.3$.

**29. B:** The technique of completing the square must be used to change:

$$4x^2 + 4y^2 - 16x - 24y + 51 = 0$$

into the standard equation of a circle. First, the constant must be moved to the right-hand side of the equals sign, and each term must be divided by the coefficient of the $x^2$ term (which is 4). The $x$ and $y$ terms must be grouped together to obtain:

$$x^2 - 4x + y^2 - 6y = -\frac{51}{4}$$

Then, the process of completing the square must be completed for each variable. This gives:

$$(x^2 - 4x + 4) + (y^2 - 6y + 9)$$

$$-\frac{51}{4} + 4 + 9$$

The equation can be written as:

$$(x - 2)^2 + (y - 3)^2 = \frac{1}{4}$$

Therefore, the center of the circle is (2, 3) and the radius is:

$$\sqrt{\frac{1}{4}} = \frac{1}{2}$$

**30. A:** Operations within the parentheses must be completed first. Then, division is completed. Finally, addition is the last operation to complete. When adding decimals, digits within each place value are added together. Therefore, the expression is evaluated as:

$$(2 \times 20) \div (7 + 1) + (6 \times 0.01) + (4 \times 0.001)$$

$$40 \div 8 + 0.06 + 0.004 = 5 + 0.06 + 0.004 = 5.064$$

**31. C:** A dollar contains 20 nickels. Therefore, if there are 12 dollars' worth of nickels, there are $12 \times 20 = 240$ nickels. Each nickel weighs 5 grams. Therefore, the weight of the nickels is $240 \times 5 = 1,200$ grams. Adding in the weight of the empty piggy bank, the filled bank weighs 2,250 grams.

**32. D:** To find Denver's total snowfall, 3 must be multiplied times $27\frac{3}{4}$. In order to easily do this, the mixed number should be converted into an improper fraction.

$$27\frac{3}{4} = \frac{27 \times 4 + 3}{4} = \frac{111}{4}$$

Therefore, Denver had approximately $\frac{3 \times 111}{4} = \frac{333}{4}$ inches of snow. The improper fraction can be converted back into a mixed number through division.

$$\frac{333}{4} = 83\frac{1}{4}\text{inches}$$

**33. D:** $x \le -5$. When solving a linear equation or inequality:

Distribution is performed if necessary: $-3(x + 4)$, or $-3x - 12 \ge x + 8$. This means that any like terms on the same side of the equation/inequality are combined.

The equation/inequality is manipulated to get the variable on one side. In this case, subtracting $x$ from both sides produces $-4x - 12 \ge 8$.

The variable is isolated using inverse operations to undo addition/subtraction. Adding 12 to both sides produces $-4x \ge 20$.

The variable is isolated using inverse operations to undo multiplication/division. Remember if dividing by a negative number, the relationship of the inequality reverses, so the sign is flipped. In this case, dividing by -4 on both sides produces $x \le -5$.

**34. C:** $y = 40x + 300$. In this scenario, the variables are the number of sales and Karen's weekly pay. The weekly pay depends on the number of sales. Therefore, weekly pay is the dependent variable ($y$), and the number of sales is the independent variable ($x$). Each pair of values from the table can be written as an ordered pair $(x, y)$: $(2, 380), (7, 580), (4, 460), (8, 620)$. The ordered pairs can be substituted into the equations to see which creates true statements (both sides equal) for each pair. Even if one ordered pair produces equal values for a given equation, the other three ordered pairs must be checked. The only equation which is true for all four ordered pairs is $y = 40x + 300$:

$$380 = 40(2) + 300 \rightarrow 380 = 380$$

$$580 = 40(7) + 300 \rightarrow 580 = 580$$

$$460 = 40(4) + 300 \rightarrow 460 = 460$$

$$620 = 40(8) + 300 \rightarrow 620 = 620$$

**35. C:** The area of the shaded region is the area of the square, minus the area of the circle. The area of the circle will be $\pi r^2$. The side of the square will be $2r$, so the area of the square will be $4r^2$. Therefore, the difference is:

$$4r^2 - \pi r^2 = (4 - \pi)r^2$$

**36. B:** The car is traveling at a speed of five meters per second. On the interval from one to three seconds, the position changes by ten meters. By making this change in position over time into a rate, the speed becomes ten meters in two seconds or five meters in one second.

**37. B:** For an ordered pair to be a solution to a system of inequalities, it must make a true statement for BOTH inequalities when substituting its values for $x$ and $y$. Substituting (-3,-2) into the inequalities produces $(-2) > 2(-3) - 3$, which is $-2 > -9$, and $(-2) < -4(-3) + 8$, or $-2 < 20$. Both are true statements.

**38. D:** The shape of the scatterplot is a parabola (U-shaped). This eliminates Choices A (a linear equation that produces a straight line) and C (an exponential equation that produces a smooth curve upward or downward). The value of $a$ for a quadratic function in standard form ($y = ax^2 + bx + c$) indicates whether the parabola opens up (U-shaped) or opens down (upside-down U). A negative value for $a$ produces a parabola that opens down; therefore, Choice B can also be eliminated.

**39. B:** According to the order of operations, multiplication and division must be completed first from left to right. Then, addition and subtraction are completed from left to right. Therefore:

$$9 \times 9 \div 9 + 9 - 9 \div 9$$

$$81 \div 9 + 9 - 9 \div 9$$

$$9 + 9 - 9 \div 9$$

$$9 + 9 - 1$$

$$18 - 1$$

$$17$$

**40. B:** First, subtract 9 from both sides to isolate the radical. Then, cube each side of the equation to obtain:

$$2x + 11 = 27$$

Subtract 11 from both sides, and then divide by 2. The result is $x = 8$. Plug 8 back into the original equation to obtain the true statement to check the answer:

$$\sqrt[3]{16 + 11} + 9 = 12$$

$$\sqrt[3]{27} + 9 = 12$$

$$3 + 9 = 12$$

**41. A:** 13 nurses. Using the given information of 1 nurse to 25 patients and 325 patients, set up an equation to solve for number of nurses ($N$):

$$\frac{N}{325} = \frac{1}{25}$$

Multiply both sides by 325 to get $N$ by itself on one side:

$$\frac{N}{1} = \frac{325}{25} = 13 \; nurses$$

**42. D:** 290 beds. Using the given information of 2 beds to 1 room and 145 rooms, set up an equation to solve for number of beds ($B$):

$$\frac{B}{145} = \frac{2}{1}$$

Multiply both sides by 145 to get $B$ by itself on one side.

$$\frac{B}{1} = \frac{290}{1} = 290 \; beds$$

**43. C:** $x = 150$. Set up the initial equation:

$$\frac{2x}{5} - 1 = 59$$

Add 1 to both sides:

$$\frac{2x}{5} - 1 + 1 = 59 + 1$$

Multiply both sides by $\frac{5}{2}$:

$$\frac{2x}{5} \times \frac{5}{2} = 60 \times \frac{5}{2} = 150$$

$$x = 150$$

**44. C:** $51.93. First, list the givens:

$$Tax = 6.0\% = 0.06$$

$$Sale = 50\% = 0.5$$

$$Hat = \$32.99$$

$$Jersey = \$64.99$$

Calculate the sale prices for hats and jerseys:

$$Hat \; sale = 0.5 \, (32.99) = 16.495$$

$$Jersey \; sale = 0.5 \, (64.99) = 32.495$$

Total the sale prices:

$$Hat \; sale + jersey \; sale = 16.495 + 32.495 = 48.99$$

Finally, calculate the sales tax and add it to the total sale prices:

$$Total\ after\ tax = 48.99 + (48.99 \times 0.06) = \$51.93$$

**45. D:** $0.45. First, list the givens:

$$Store\ coffee = \$1.23/lb$$

$$Local\ roaster\ coffee = \$1.98/1.5\ lb$$

Calculate the cost for 5 pounds of store brand coffee.

$$\frac{\$1.23}{1\ lb} \times 5\ lb = \$6.15$$

Calculate the cost for 5 pounds of the local roaster's coffee:

$$\frac{\$1.98}{1.5\ lb} \times 5\ lb = \$6.60$$

Subtract to find the difference in price for 5 pounds of coffee:

$$\$6.60 - \$6.15 = \$0.45$$

**46. D:** $3,325. First, list the givens:

$$1,800\ ft = \$2,000$$

$$Cost\ after\ 1,800\ ft = \$1.00/ft$$

Find how many feet left after the first 1,800 ft:

$$3125\ ft - 1,800\ ft = 1325\ ft$$

Calculate the cost for the feet over 1,800 ft:

$$1,325\ ft \times \frac{\$1.00}{1\ ft} = \$1,325$$

Total for entire cost:

$$\$2,000 + \$1,325 = \$3,325$$

**47. A:** 12. Calculate how many gallons the bucket holds:

$$11.4\ L \times \frac{1\ gal}{3.8\ L} = 3\ gal$$

Now how many buckets to fill the pool which needs 35 gallons:

$$\frac{35\ gal}{3\ gal} = 11.67$$

Since the amount is more than 11 but less than 12, we must fill the bucket 12 times.

**48. D:** Three girls for every two boys can be expressed as a ratio: 3:2. This can be visualized as splitting the school into 5 groups: 3 girl groups and 2 boy groups. The number of students in each group can be found by dividing the total number of students by 5:

$$\frac{650 \text{ students}}{5 \text{ groups}} = \frac{130 \text{ students}}{\text{group}}$$

To find the total number of girls, multiply the number of students per group (130) by the number of girl groups in the school (3). This equals 390, Choice *D*.

**49. C:** The volume of a pyramid is $length \times width \times height$, divided by 3, and $6 \times 6 \times 9$, divided by 3 is 108 in³. Choice *A* is incorrect because 324 in³ is $length \times width \times height$ without dividing by 3. Choice *B* is incorrect because 6 is used for height instead of 9 ($6 \times 6 \times 6$) and divided by 3 to get 72 in³. Choice *D* is incorrect because 18 in³ is $6 \times 9$, divided by 3, and leaving out a 6.

**50. A:** 22%. Converting from a fraction to a percentage generally involves two steps. First, the fraction needs to be converted to a decimal:

$$\frac{2}{9} = 0.\overline{22}$$

The top line indicates that the decimal actually goes on forever with an endless amount of 2's.

Second, the decimal needs to be moved two places to the right to convert to a percent:

$$22\%$$

**51. A:** The volume of a cone is $\pi r^2 h$, divided by 3, and $\pi \times 10^2 \times 12$, divided by 3, is 400 cm³. Choice *B* is $10^2 \times 2$. Choice *C* is incorrect because it is $10 \times 12$. Choice *D* is also incorrect because that is $10^2 + 40$.

**52. A:** 37.5%. Solve this by setting up the percent formula:

$$\frac{3}{8} = \frac{\%}{100}$$

Multiply 3 by 100 to get 300. Then divide 300 by 8:

$$\frac{300}{8} = 37.5\%$$

Note that with the percent formula, 37.5 is automatically a percentage and does not need to have any further conversions.

**53. C:** 216 cm. Because area is a two-dimensional measurement, the dimensions are multiplied by a scale that is squared to determine the scale of the corresponding areas. The dimensions of the rectangle are multiplied by a scale of 3. Therefore, the area is multiplied by a scale of $3^2$ (which is equal to 9):

$$24 \; cm \times 9 = 216 \; cm$$

**54.** Add 3 to both sides to get $4x = 8$. Then, divide both sides by 4 to get $x = 2$.

**55.** To solve this correctly, keep in mind the order of operations with the mnemonic PEMDAS (Please Excuse My Dear Aunt Sally). This stands for Parentheses, Exponents, Multiplication, Division, Addition, Subtraction. Taking it step by step, solve inside the parentheses first:

$$4 \times 7 + (4)^2 \div 2$$

Then, apply the exponent:

$$4 \times 7 + 16 \div 2$$

Multiplication and division are both performed next:

$$28 + 8 = 36$$

Addition and subtraction are done last. The solution is 36.

**56.** Follow the order of operations in order to solve this problem. Simplify the radicals within the parentheses first, and then follow the remainder as usual:

$$(6 \times 4) - 9$$

This equals $24 - 9$, or 15.

**57.** For an even number of total values, the *median* is calculated by finding the *mean* or average of the two middle values once all values have been arranged in ascending order from least to greatest. In this case, $(92 + 83) \div 2$ would equal the median 87.5.

**58.** The formula for the perimeter of a rectangle is $P = 2L + 2W$, where $P$ is the perimeter, $L$ is the length, and $W$ is the width. The first step is to substitute all of the data into the formula:

$$36 = 2(12) + 2W$$

Simplify by multiplying $2 \times 12$:

$$36 = 24 + 2W$$

Simplifying this further by subtracting 24 on each side, which gives:

$$36 - 24 = 24 - 24 + 2W$$

$$12 = 2W$$

Divide by 2:

$$6 = W$$

The width is 6 cm. Remember to test this answer by substituting this value into the original formula:

$$36 = 2(12) + 2(6)$$

# SAT Math Practice Test #2

1. If a car can travel 300 miles in 4 hours, how far can it go in an hour and a half?
    a. 100 miles
    b. 112.5 miles
    c. 135.5 miles
    d. 150 miles

2. At the store, Jan spends $90 on apples and oranges. Apples cost $1 each and oranges cost $2 each. If Jan buys the same number of apples as oranges, how many oranges did she buy?
    a. 20
    b. 25
    c. 30
    d. 35

3. What is the volume of a box with rectangular sides 5 feet long, 6 feet wide, and 3 feet high?
    a. 60 cubic feet
    b. 75 cubic feet
    c. 90 cubic feet
    d. 14 cubic feet

4. A train traveling 50 miles per hour takes a trip lasting 3 hours. If a map has a scale of 1 inch per 10 miles, how many inches apart are the train's starting point and ending point on the map if it travelled in a straight line?
    a. 14
    b. 12
    c. 13
    d. 15

5. A traveler takes an hour to drive to a museum, spends 3 hours and 30 minutes there, and takes half an hour to drive home. What percentage of his or her time was spent driving?
    a. 15%
    b. 30%
    c. 40%
    d. 60%

6. A truck is carrying three cylindrical barrels. Their bases have a diameter of 2 feet, and they have a height of 3 feet. What is the total volume of the three barrels in cubic feet?
    a. $3\pi$
    b. $9\pi$
    c. $12\pi$
    d. $15\pi$

7. Greg buys a $10 lunch with 5% sales tax. He leaves a $2 tip after his bill. How much money does he spend?
    a. $12.50
    b. $12
    c. $13
    d. $13.25

8. Marty wishes to save $150 over a 4-day period. How much must Marty save each day on average?
    a. $37.50
    b. $35
    c. $45.50
    d. $41

9. Bernard can make $80 per day. If he needs to make $300 and only works full days, how many days will this take?
    a. 6
    b. 3
    c. 5
    d. 4

10. A couple buys a house for $150,000. They sell it for $165,000. By what percentage did the house's value increase?
    a. 10%
    b. 13%
    c. 15%
    d. 17%

11. A school has 15 teachers and 20 teaching assistants. They have 200 students. What is the ratio of faculty to students?
    a. 3:20
    b. 4:17
    c. 5:54
    d. 7:40

12. A map has a scale of 1 inch per 5 miles. A car can travel 60 miles per hour. If the distance from the start to the destination is 3 inches on the map, how long will it take the car to make the trip?
    a. 12 minutes
    b. 15 minutes
    c. 17 minutes
    d. 20 minutes

13. Taylor works two jobs. The first pays $20,000 per year. The second pays $10,000 per year. She donates 15% of her income to charity. How much does she donate each year?
    a. $4500
    b. $5000
    c. $5500
    d. $6000

14. A box with rectangular sides is 24 inches wide, 18 inches deep, and 12 inches high. What is the volume of the box in cubic feet?

    a. 2
    b. 3
    c. 4
    d. 5

15. Kristen purchases $100 worth of CDs and DVDs. The CDs cost $10 each and the DVDs cost $15. If she bought four DVDs, how many CDs did she buy?

    a. 5
    b. 6
    c. 3
    d. 4

16. If Sarah reads at an average rate of 21 pages in four nights, how long will it take her to read 140 pages?

    a. 6 nights
    b. 26 nights
    c. 8 nights
    d. 27 nights

17. Mom's car drove 72 miles in 90 minutes. There are 5280 feet per mile. How fast did she drive in feet per second?

    a. 0.8 feet per second
    b. 48.9 feet per second
    c. 0.009 feet per second
    d. 70.4 feet per second

18. This chart indicates how many sales of CDs, vinyl records, and MP3 downloads occurred over the last year. Approximately what percentage of the total sales was from CDs?

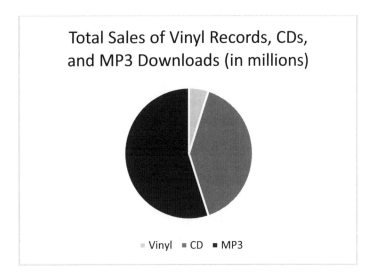

Total Sales of Vinyl Records, CDs, and MP3 Downloads (in millions)

▪ Vinyl ▪ CD ▪ MP3

a. 55%
b. 25%
c. 40%
d. 5%

19. After a 20% sale discount, Frank purchased a new refrigerator for $850. How much did he save from the original price?
a. $170
b. $212.50
c. $105.75
d. $200

20. Which of the following is NOT a way to write 40 percent of $N$?
a. $(0.4)N$

b. $\frac{2}{5}N$

c. $40N$

d. $\frac{4N}{10}$

21. The graph of which function has an $x$-intercept of $-2$?
a. $y = 2x - 3$
b. $y = 4x + 2$
c. $y = x^2 + 5x + 6$
d. $y = -\frac{1}{2} \times 2^x$

22. The table below displays the number of three-year-olds at Kids First Daycare who are potty-trained and those who still wear diapers.

|  | Potty-trained | Wear diapers | Sum |
|---|---|---|---|
| **Boys** | 26 | 22 | 48 |
| **Girls** | 34 | 18 | 52 |
| **Total** | 60 | 40 |  |

What is the probability that a three-year-old girl chosen at random from the school is potty-trained?
 a. 52 percent
 b. 34 percent
 c. 65 percent
 d. 57 percent

23. A clothing company with a target market of U.S. boys surveys 2,000 twelve-year-old boys to find their height. The average height of the boys is 61 inches. For the above scenario, 61 inches represents which of the following?
 a. Sample statistic
 b. Population parameter
 c. Confidence interval
 d. Measurement error

24. A government agency is researching the average consumer cost of gasoline throughout the United States. Which data collection method would produce the most valid results?
 a. Randomly choosing one hundred gas stations in the state of New York
 b. Randomly choosing ten gas stations from each of the fifty states
 c. Randomly choosing five hundred gas stations from across all fifty states with the number chosen proportional to the population of the state
 d. All three methods would each produce equally valid results.

25. Suppose an investor deposits $1,200 into a bank account that accrues 1 percent interest per month. Assuming $x$ represents the number of months since the deposit and $y$ represents the money in the account, which of the following exponential functions models the scenario?
 a. $y = (0.01)(1200^x)$
 b. $y = (1200)(0.01^x)$
 c. $y = (1.01)(1200^x)$
 d. $y = (1200)(1.01^x)$

26. A student gets an 85% on a test with 20 questions. How many answers did the student solve correctly?
 a. 15
 b. 16
 c. 17
 d. 18

27. Four people split a bill. The first person pays for $\frac{1}{5}$, the second person pays for $\frac{1}{4}$, and the third person pays for $\frac{1}{3}$. What fraction of the bill does the fourth person pay?

    a. $\frac{13}{60}$

    b. $\frac{47}{60}$

    c. $\frac{1}{4}$

    d. $\frac{4}{15}$

28. Which of the following fractions is equal to 9.3?

    a. $9\frac{3}{7}$

    b. $\frac{903}{1000}$

    c. $\frac{9.03}{100}$

    d. $9\frac{3}{10}$

29. What is the solution to $3\frac{2}{3} - 1\frac{4}{5}$?

    a. $1\frac{13}{15}$

    b. $\frac{14}{15}$

    c. $2\frac{2}{3}$

    d. $\frac{4}{5}$

30. What is $\frac{420}{98}$ rounded to the nearest integer?

    a. 4

    b. 3

    c. 5

    d. 6

31. What is the solution to $4\frac{1}{3} + 3\frac{3}{4}$?

    a. $6\frac{5}{12}$

    b. $8\frac{1}{12}$

    c. $8\frac{2}{3}$

    d. $7\frac{7}{12}$

32. Five of six numbers have a sum of 25. The average of all six numbers is 6. What is the sixth number?
    a. 8
    b. 10
    c. 11
    d. 12

33. Suppose the function $y = \frac{1}{8}x^3 + 2x - 21$ approximates the population of a given city between the years 1900 and 2000 with $x$ representing the year ($1900 = 0$) and $y$ representing the population (in 1000s). Which of the following domains are relevant for the scenario?
    a. $(-\infty, \infty)$
    b. $[1900, 2000]$
    c. $[0, 100]$
    d. $[0, 0]$

34. What is the equation of a circle whose center is (0, 0) and whole radius is 5?
    a. $(x - 5)^2 + (y - 5)^2 = 25$
    b. $(x)^2 + (y)^2 = 5$
    c. $(x)^2 + (y)^2 = 25$
    d. $(x + 5)^2 + (y + 5)^2 = 25$

35. What is the equation of a circle whose center is (1, 5) and whole radius is 4?
    a. $(x - 1)^2 + (y - 25)^2 = 4$
    b. $(x - 1)^2 + (y - 25)^2 = 16$
    c. $(x + 1)^2 + (y + 5)^2 = 16$
    d. $(x - 1)^2 + (y - 5)^2 = 16$

36. Where does the point (-3, -4) lie on the circle with the equation $(x)^2 + (y)^2 = 25$?
    a. Inside of the circle.
    b. Outside of the circle.
    c. On the circle.
    d. There is not enough information to tell.

37. A drug needs to be stored at room temperature (68 °F). What is the equivalent temperature in degrees Celsius?
    a. 36 °C
    b. 72 °C
    c. 68 °C
    d. 20 °C

38. What is the slope of this line?

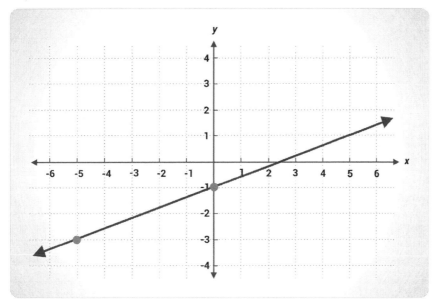

a. 2

b. $\frac{5}{2}$

c. $\frac{1}{2}$

d. $\frac{2}{5}$

39. What is the perimeter of the figure below? Note that the solid outer line is the perimeter.

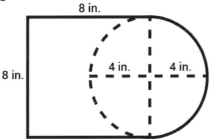

a. 48.566 ft
b. 36.566 ft
c. 19.78 ft
d. 30.566 ft

40. Which of the following equations best represents the problem below?
The width of a rectangle is 2 centimeters less than the length. If the perimeter of the rectangle is 44 centimeters, then what are the dimensions of the rectangle?

a. $2l + 2(l - 2) = 44$
b. $(l + 2) + (l + 2) + l = 48$
c. $l \times (l - 2) = 44$
d. $(l + 2) + (l + 2) + l = 44$

41. How will the following algebraic expression be simplified: $(5x^2 - 3x + 4) - (2x^2 - 7)$?
    a. $x^5$
    b. $3x^2 - 3x + 11$
    c. $3x^2 - 3x - 3$
    d. $x - 3$

42. What is 39% of 164?
    a. 63.96%
    b. 23.78%
    c. 6,396%
    d. 2.38%

43. Kimberley earns $10 an hour babysitting, and after 10 p.m., she earns $12 an hour, with the amount paid being rounded to the nearest hour accordingly. On her last job, she worked from 5:30 p.m. to 11 p.m. In total, how much did Kimberley earn for that job?
    a. $45
    b. $57
    c. $62
    d. $42

44. Keith's bakery had 252 customers go through its doors last week. This week, that number increased to 378. By what percentage did his customer volume increase?
    a. 26%
    b. 50%
    c. 35%
    d. 12%

## No Calculator Questions

45. A family purchases a vehicle in 2005 for $20,000. In 2010, they decide to sell it for a newer model. They are able to sell the car for $8,000. By what percentage did the value of the family's car drop?
    a. 40%
    b. 68%
    c. 60%
    d. 33%

46. In May of 2010, a couple purchased a house for $100,000. In September of 2016, the couple sold the house for $93,000 so they could purchase a bigger one to start a family. How many months did they own the house?
    a. 76
    b. 54
    c. 85
    d. 93

47. At the beginning of the day, Xavier has 20 apples. At lunch, he meets his sister Emma and gives her half of his apples. After lunch, he stops by his neighbor Jim's house and gives him 6 of his apples. He then uses 3/4 of his remaining apples to make an apple pie for dessert at dinner. At the end of the day, how many apples does Xavier have left?

    a. 4
    b. 6
    c. 2
    d. 1

48. If $\frac{5}{2} \div \frac{1}{3} = n$, then $n$ is between:

    a. 5 and 7
    b. 7 and 9
    c. 9 and 11
    d. 3 and 5

49. A closet is filled with red, blue, and green shirts. If $\frac{1}{3}$ of the shirts are green and $\frac{2}{5}$ are red, what fraction of the shirts are blue?

    a. $\frac{4}{15}$

    b. $\frac{1}{5}$

    c. $\frac{7}{15}$

    d. $\frac{1}{2}$

50. Shawna buys $2\frac{1}{2}$ gallons of paint. If she uses $\frac{1}{3}$ of it on the first day, how much does she have left?

    a. $1\frac{5}{6}$ gallons

    b. $1\frac{1}{2}$ gallons

    c. $1\frac{2}{3}$ gallons

    d. 2 gallons

51. What is the volume of a cylinder, in terms of $\pi$, with a radius of 6 centimeters and a height of 2 centimeters?

    a. $36\,\pi$ cm$^3$
    b. $24\,\pi$ cm$^3$
    c. $72\,\pi$ cm$^3$
    d. $48\,\pi$ cm$^3$

52. What is the length of the hypotenuse of a right triangle with one leg equal to 3 centimeters and the other leg equal to 4 centimeters?

    a. 7 cm
    b. 5 cm
    c. 25 cm
    d. 12 cm

53. Twenty is 40% of what number?
  a. 50
  b. 8
  c. 200
  d. 5000

54. If Danny takes 48 minutes to walk 3 miles, how many minutes should it take him to walk 5 miles maintaining the same speed?

55. The perimeter of a 6-sided polygon is 56 cm. The lengths of three sides are 9 cm each. The lengths of two other sides are 8 cm each. What is the length of the missing side?

56. If the sine of $30° = x$, the cosine of what angle, in degrees, also equals $x$?

57. What is the value of $x^2 - 2xy + 2y^2$ when $x = 2$ and $y = 3$?

58. What is the value of $x$ if $4x - 3 = 5$?

# Answer Explanations #2

**1. B:** 300 miles in 4 hours is $\frac{300}{4} = 75$ miles per hour. In 1.5 hours, the car will go $1.5 \times 75$ miles, or 112.5 miles.

**2. C:** One apple/orange pair costs $3 total. Therefore, Jan bought $\frac{90}{3} = 30$ total pairs, and hence, she bought 30 oranges.

**3. C:** The formula for the volume of a box with rectangular sides is the length times width times height, so $5 \times 6 \times 3 = 90$ cubic feet.

**4. D:** First, the train's journey in the real word is $3 \times 50 = 150$ miles. On the map, 1 inch corresponds to 10 miles, so there is $\frac{150}{10} = 15$ inches on the map.

**5. B:** The total trip time is $1 + 3.5 + 0.5 = 5$ hours. The total time driving is $1 + 0.5 = 1.5$ hours. So, the fraction of time spent driving is 1.5 / 5 or 3 / 10.

To get the percentage, convert this to a fraction out of 100. The numerator and denominator are multiplied by 10, with a result of 30 / 100. The percentage is the numerator in a fraction out of 100, so 30%.

**6. B:** The formula for the volume of a cylinder is $\pi r^2 h$, where $r$ is the radius and $h$ is the height. The diameter is twice the radius, so these barrels have a radius of 1 foot. That means each barrel has a volume of:

$$\pi \times 1^2 \times 3 = 3\pi \text{ cubic feet}$$

Since there are three of them, the total is $3 \times 3\pi = 9\pi$ cubic feet.

**7. A:** The tip is not taxed, so he pays 5% tax only on the $10. The tax is 5% of $10, or $0.05 \times 10 = \$0.50$. Add up $10 + $2 + $0.50 to get $12.50.

**8. A:** The first step is to divide up $150 into four equal parts. $150 \div 4$ is 37.5, so she needs to save an average of $37.50 per day.

**9. D:** The number of days can be found by taking the total amount Bernard needs to make and dividing it by the amount he earns per day:

$$\frac{300}{80} = \frac{30}{8} = \frac{15}{4} = 3.75$$

But Bernard is only working full days, so he will need to work 4 days, since 3 days is not a sufficient amount of time.

**10. A:** The value went up by $165,000 - $150,000 = $15,000.

Out of $150,000, this is $\frac{15,000}{150,000} = \frac{1}{10}$.

Convert this to having a denominator of 100, which yields a result of $\frac{10}{100}$ or 10%.

**11. D:** The total faculty is $15 + 20 = 35$. Therefore, the faculty to student ratio is 35:200. Then, to simplify this ratio, both the numerator and the denominator are divided by 5, since 5 is a common factor of both, which yields 7:40.

**12. B:** The journey will be $5 \times 3 = 15$ miles. A car traveling at 60 miles per hour is traveling at 1 mile per minute. The resulting equation would be:

$$\frac{15 \text{ mi}}{1 \frac{\text{mi}}{\text{min}}} = 15 \text{ min}$$

Therefore, it will take 15 minutes to make the journey.

**13. A:** Taylor's total income is $\$20,000 + \$10,000 = \$30,000$. Fifteen percent of this is $\frac{15}{100} = \frac{3}{20}$. So:

$$\frac{3}{20} \times \$30,000 = \frac{\$90,000}{20}$$

$$\frac{\$9000}{2} = \$4500$$

**14. B:** Since the answer will be in cubic feet rather than inches, the first step is to convert from inches to feet for the dimensions of the box. There are 12 inches per foot, so the box is $24 \div 12 = 2$ feet wide, $18 \div 12 = 1.5$ feet deep, and $12 \div 12 = 1$ foot high. The volume is the product of these three together:

$$2 \times 1.5 \times 1 = 3 \text{ cubic feet}$$

**15. D:** Kristen bought four DVDs, which would cost a total of $4 \times 15 = \$60$. She spent a total of $100, so she spent $\$100 - \$60 = \$40$ on CDs. Since they cost $10 each, she must have purchased $40 \div 10 = 4$ CDs.

**16. D:** This problem can be solved by setting up a proportion involving the given information and the unknown value. The proportion is:

$$\frac{21 \text{ pages}}{4 \text{ nights}} = \frac{140 \text{ pages}}{x \text{ nights}}$$

Solving the proportion by cross-multiplying, the equation becomes $21x = 4 \times 140$, where $x = 26.67$. Since it is not an exact number of nights, the answer is rounded up to 27 nights. Twenty-six nights would not give Sarah enough time.

**17. D:** This problem can be solved by using unit conversion. The initial units are miles per minute. The final units need to be feet per second. Converting miles to feet uses the equivalence statement 1 mile equals 5,280 feet. Converting minutes to seconds uses the equivalence statement 1 minute equals 60 seconds. Setting up the ratios to convert the units is shown in the following equation:

$$\frac{72 \text{ mi}}{90 \text{ min}} \times \frac{1 \text{ min}}{60 \text{ s}} \times \frac{5280 \text{ ft}}{1 \text{ mi}} = 70.4 \frac{\text{ft}}{\text{s}}$$

The initial units cancel out, and the new units are left.

**18. C:** The sum total percentage of a pie chart must equal 100%. Since the CD sales take up less than half of the chart and more than a quarter (25%), it can be determined to be 40% overall. This can also be measured with a protractor. The angle of a circle is 360°. Since 25% of 360° would be 90° and 50% would be 180°, the angle percentage of CD sales falls in between; therefore, it would be Choice C.

**19. B:** Since $850 is the price *after* a 20% discount, $850 represents 80% of the original price. To determine the original price, set up a proportion with the ratio of the sale price (850) to original price (unknown) equal to the ratio of sale percentage (where $x$ represents the unknown original price):

$$\frac{850}{x} = \frac{80}{100}$$

To solve a proportion, cross multiply the numerators and denominators and set the products equal to each other:

$$(850)(100) = (80)(x)$$

Multiplying each side results in the equation $85,000 = 80x$.

To solve for $x$, divide both sides by 80: $\frac{85,000}{80} = \frac{80x}{80}$, resulting in $x = 1062.5$. Remember that $x$ represents the original price. Subtracting the sale price from the original price ($1062.50 - $850) indicates that Frank saved $212.50.

**20. C:** $40N$ would be 4000% of $N$. It's possible to check that each of the others is actually 40% of $N$.

**21. C:** An $x$-intercept is the point where the graph crosses the $x$-axis. At this point, the value of $y$ is 0. To determine if an equation has an $x$-intercept of -2, substitute -2 for $x$, and calculate the value of $y$. If the value of -2 for $x$ corresponds with a $y$-value of 0, then the equation has an $x$-intercept of -2. The only answer choice that produces this result is Choice C:

$$0 = (-2)^2 + 5(-2) + 6$$

**22. C:** The conditional frequency of a girl being potty-trained is calculated by dividing the number of potty-trained girls by the total number of girls: $34 \div 52 = 0.65$. To determine the conditional probability, multiply the conditional frequency by 100: $0.65 \times 100 = 65\%$.

**23. A:** A sample statistic indicates information about the data that was collected (in this case, the heights of those surveyed). A population parameter describes an aspect of the entire population (in this case, all twelve-year-old boys in the United States). A confidence interval would consist of a range of heights likely to include the actual population parameter. Measurement error relates to the validity of the data that was collected.

**24. C:** To ensure valid results, samples should be taken across the entire scope of the study. Since all states are not equally populated, representing each state proportionately would result in a more accurate statistic.

**25. D:** Exponential functions can be written in the form: $y = a \times b^x$. The equation for an exponential function can be written given the $y$-intercept ($a$) and the growth rate ($b$).

The *y*-intercept is the output (*y*) when the input (*x*) equals zero. It can be thought of as an "original value," or starting point. The value of *b* is the rate at which the original value increases (*b* > 1) or decreases (*b* < 1).

In this scenario, the *y*-intercept, *a*, would be $1200, and the growth rate, *b*, would be 1.01 (100% of the original value combined with 1% interest, or 100% + 1% = 101% = 1.01).

**26. C:** 85% of a number means multiplying that number by 0.85. So, $0.85 \times 20 = \frac{85}{100} \times \frac{20}{1}$, which can be simplified to:

$$\frac{17}{20} \times \frac{20}{1} = 17$$

**27. A:** To find the fraction of the bill that the first three people pay, the fractions need to be added, which means finding the common denominator. The common denominator will be 60:

$$\frac{1}{5} + \frac{1}{4} + \frac{1}{3} = \frac{12}{60} + \frac{15}{60} + \frac{20}{60} = \frac{47}{60}$$

The remainder of the bill is:

$$1 - \frac{47}{60} = \frac{60}{60} - \frac{47}{60} = \frac{13}{60}$$

**28. D:** $9\frac{3}{10}$. To convert a decimal to a fraction, remember that any number to the left of the decimal point will be a whole number. Then, sense 0.3 goes to the tenths place, it can be placed over 10.

**29. A:** Convert the mixed fractions to improper fractions: $\frac{11}{3} - \frac{9}{5}$. Subtract using 15 as a common denominator and rewrite to get rid of the improper fraction:

$$\frac{11}{3} - \frac{9}{5} = \frac{55}{15} - \frac{27}{15} = \frac{28}{15} = 1\frac{13}{15}$$

**30. A:** Dividing by 98 can be approximated by dividing by 100, which would mean shifting the decimal point of the numerator to the left by 2. The result is 4.2 and rounds to 4.

**31. B:** First, separate out and add the whole numbers from the mixed fractions:

$$4\frac{1}{3} + 3\frac{3}{4}$$

$$4 + 3 + \frac{1}{3} + \frac{3}{4}$$

$$7 + \frac{1}{3} + \frac{3}{4}$$

Adding the fractions gives:

$$\frac{1}{3} + \frac{3}{4}$$

$$\frac{4}{12} + \frac{9}{12}$$

$$\frac{13}{12}$$

$$1 + \frac{1}{12}$$

Thus:

$$7 + \frac{1}{3} + \frac{3}{4} = 7 + 1 + \frac{1}{12} = 8\frac{1}{12}$$

**32. C:** The average is calculated by adding all six numbers, then dividing by 6. The first five numbers have a sum of 25. If the total divided by 6 is equal to 6, then the total itself must be 36. The sixth number must be $36 - 25 = 11$.

**33. C:** The domain consists of all possible inputs, or $x$-values. The scenario states that the function approximates the population between the years 1900 and 2000. It also states that $x = 0$ represents the year 1900. Therefore, the year 2000 would be represented by $x = 100$. Only inputs between 0 and 100 are relevant in this case.

**34. C:** Nothing is added to $x$ and $y$ since the center is 0 and $5^2$ is 25. Choice *A* is not the correct answer because you do not subtract the radius from $x$ and $y$. Choice *B* is not the correct answer because you must square the radius on the right side of the equation. Choice *D* is not the correct answer because you do not add the radius to $x$ and $y$ in the equation.

**35. D:** Subtract the center from the $x$ and $y$ values of the equation and square the radius on the right side of the equation. Choice *A* is not the correct answer because you need to square the radius of the equation. Choice *B* is not the correct answer because you do not square the centers of the equation. Choice *C* is not the correct answer because you need to subtract (not add) the centers of the equation.

**36. C:** Plug in the values for $x$ and $y$ to discover that the solution works, which is:

$$(-3)^2 + (-4)^2 = 25$$

Choices *A* and *B* are not the correct answers since the solution works. Choice *D* is not the correct answer because there is enough information to tell where the given point lies on the circle.

**37. D:** The correct answer of 20 °C can be found using the appropriate temperature conversion formula:

$$°C = (°F - 32) \times \frac{5}{9}$$

**38. D:** The slope is given by the change in $y$ divided by the change in $x$. Specifically, it's:

$$slope = \frac{y_2 - y_1}{x_2 - x_1}$$

The first point is (-5, -3), and the second point is (0, -1). Work from left to right when identifying coordinates. Thus the point on the left is point 1 (-5,-3) and the point on the right is point 2 (0,-1).

Now we need to just plug those numbers into the equation:

$$slope = \frac{-1 - (-3)}{0 - (-5)}$$

It can be simplified to:

$$slope = \frac{-1 + 3}{0 + 5}$$

$$slope = \frac{2}{5}$$

**39. B:** The figure is composed of three sides of a square and a semicircle. The sides of the square are simply added: $8 + 8 + 8 = 24$ feet. The circumference of a circle is found by the equation $C = 2\pi r$. The radius is 4, so the circumference of the circle is 25.132 feet. Only half of the circle makes up the outer border of the figure (part of the perimeter) so half of 25.132 feet is 12.566 feet. Therefore, the total perimeter is: $24 \text{ ft} + 12.566 \text{ ft} = 36.566 \text{ ft}$. The other answer choices use the incorrect formula or fail to include all of the necessary sides.

**40. A:** The first step is to determine the unknown, which is in terms of the length, $l$.

The second step is to translate the problem into the equation using the perimeter of a rectangle:

$$P = 2l + 2w$$

The width is the length minus 2 centimeters. The resulting equation is:

$$2l + 2(l - 2) = 44$$

The equation can be solved as follows:

| | |
|---|---|
| $2l + 2l - 4 = 44$ | Apply the distributive property on the left side of the equation |
| $4l - 4 = 44$ | Combine like terms on the left side of the equation |
| $4l = 48$ | Add 4 to both sides of the equation |
| $l = 12$ | Divide both sides of the equation by 4 |

The length of the rectangle is 12 centimeters. The width is the length minus 2 centimeters, which is 10 centimeters. Checking the answers for length and width forms the following equation:

$$44 = 2(12) + 2(10)$$

The equation can be solved using the order of operations to form a true statement: $44 = 44$.

**41. B:** $3x^2 - 3x + 11$. By distributing the implied one in front of the first set of parentheses and the -1 in front of the second set of parentheses, the parenthesis can be eliminated:

$$1(5x^2 - 3x + 4) - 1(2x^2 - 7)$$

$$5x^2 - 3x + 4 - 2x^2 + 7$$

Next, like terms (same variables with same exponents) are combined by adding the coefficients and keeping the variables and their powers the same:

$$5x^2 - 3x + 4 - 2x^2 + 7 = 3x^2 - 3x + 11$$

**42. A:** This question involves the percent formula. Since we're beginning with a percent, also known as a number over 100, we'll put 39 on the right side of the equation:

$$\frac{x}{164} = \frac{39}{100}$$

Now, multiply 164 and 39 to get 6,396, which then needs to be divided by 100.

$$6,396 \div 100 = 63.96$$

**43. C:** Kimberley worked 4.5 hours at the rate of $10/h and 1 hour at the rate of $12/h. The problem states that her pay is rounded to the nearest hour, so the 4.5 hours would round up to 5 hours at the rate of $10/h.

$$(5h) \times \left(\frac{\$10}{h}\right) + (1h) \times \left(\frac{\$12}{h}\right)$$

$$\$50 + \$12 = \$62$$

**44. B:** The first step is to calculate the difference between the larger value and the smaller value:

$$378 - 252 = 126$$

To calculate this difference as a percentage of the original value, and thus calculate the percentage *increase*, 126 is divided by 252, then this result is multiplied by 100 to find the percentage: 50%, or Choice *B*.

**45. C:** In order to find the percentage by which the value of the car has been reduced, the current cash value should be subtracted from the initial value and then the difference divided by the initial value. The result should be multiplied by 100 to find the percentage decrease.

$$\frac{20,000 - 8,000}{20,000} = 0.6$$

$$(0.6) \times 100 = 60\%$$

**46. A:** This problem can be solved by simple multiplication and addition. Since the sale date is over six years apart, 6 can be multiplied by 12 for the number of months in a year, and then the remaining 4 months can be added.

$$(6 \times 12) + 4 = ?$$

$$72 + 4 = 76$$

**47. D:** This problem can be solved using basic arithmetic. Xavier starts with 20 apples, then gives his sister half, so 20 divided by 2.

$$\frac{20}{2} = 10$$

He then gives his neighbor 6, so 6 is subtracted from 10.

$$10 - 6 = 4$$

Lastly, he uses 3/4 of his apples to make an apple pie, so to find remaining apples, the first step is to subtract 3/4 from one and then multiply the difference by 4.

$$\left(1 - \frac{3}{4}\right) \times 4 = ?$$

$$\left(\frac{4}{4} - \frac{3}{4}\right) \times 4 = ?$$

$$\left(\frac{1}{4}\right) \times 4 = 1$$

**48. B:** To find the value of $n$, multiply $\frac{5}{2}$ by the reciprocal of $\frac{1}{3}$ and simplify:

$$n = \frac{5}{2} \div \frac{1}{3} = \frac{5}{2} \times \frac{3}{1} = \frac{15}{2} = 7.5$$

**49. A:** The total fraction taken up by green and red shirts will be:

$$\frac{1}{3} + \frac{2}{5} = \frac{5}{15} + \frac{6}{15} = \frac{11}{15}$$

The remaining fraction is:

$$1 - \frac{11}{15} = \frac{15}{15} - \frac{11}{15} = \frac{4}{15}$$

**50. C:** If she has used 1/3 of the paint, she has 2/3 remaining. $2\frac{1}{2}$ gallons are the same as $\frac{5}{2}$ gallons. The calculation is:

$$\frac{2}{3} \times \frac{5}{2} = \frac{5}{3} = 1\frac{2}{3} \; gallons$$

**51. C:** The volume of a cylinder is $\pi r^2 h$, and $\pi \times 6^2 \times 2$ is $72\,\pi$ cm$^3$. Choice *A* is not the correct answer because that is only $6^2 \times \pi$. Choice *B* is not the correct answer because that is $2^2 \times 6 \times \pi$. Choice *D* is not the correct answer because that is $2^3 \times 6 \times \pi$.

**52. B:** This answer is correct because $3^2 + 4^2$ is $9 + 16$, which is 25. Taking the square root of 25 is 5. Choice *A* is not the correct answer because that is $3 + 4$. Choice *C* is not the correct answer because that is stopping at $3^2 + 4^2$ is $9 + 16$, which is 25. Choice *D* is not the correct answer because that is $3 \times 4$.

**53. A:** Setting up a proportion is the easiest way to represent this situation. The proportion becomes $\frac{20}{x} = \frac{40}{100}$, where cross-multiplication can be used to solve for $x$. Here, $40x = 2000$, so $x = 50$.

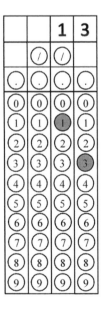

**54.** To solve the problem, a proportion is written consisting of ratios comparing distance and time. One way to set up the proportion is $\frac{3}{48} = \frac{5}{x}$ or $\left(\frac{distance}{time} = \frac{distance}{time}\right)$, where $x$ represents the unknown value of time. To solve a proportion, the ratios are cross-multiplied:

$$(3)(x) = (5)(48)$$

$$3x = 240$$

The equation is solved by isolating the variable, or dividing by 3 on both sides, to produce $x = 80$.

**55.** Perimeter is found by calculating the sum of all sides of the polygon. $9 + 9 + 9 + 8 + 8 + s = 56$, where $s$ is the missing side length. Therefore, 43 plus the missing side length is equal to 56. The missing side length is 13 cm.

**56.** When $x$ and $y$ are complementary angles, the sine of $x$ is equal to the cosine of $y$. The complementary angle of 30 is $90 - 30 = 60$ degrees. Therefore, the answer is 60 degrees.

**57.** Start with the original equation: $x - 2xy + 2y$, then replace each instance of $x$ with a 2, and each instance of $y$ with a 3 to get:

$$2^2 - 2 \times 2 \times 3 + 2 \times 3^2$$

$$4 - 12 + 18 = 10$$

58. Add 3 to both sides to get $4x = 8$. Then divide both sides by 4 to get $x = 2$

# SAT Math Practice Test #3

1. If $6t + 4 = 16$, what is $t$?
   a. 1
   b. 2
   c. 3
   d. 4

2. The variable $y$ is directly proportional to $x$. If $y = 3$ when $x = 5$, then what is $y$ when $x = 20$?
   a. 10
   b. 12
   c. 14
   d. 16

3. A line passes through the point (1, 2) and crosses the $y$-axis at $y = 1$. Which of the following is an equation for this line?
   a. $y = 2x$
   b. $y = x + 1$
   c. $x + y = 1$
   d. $y = \frac{x}{2} - 2$

4. There are $4x + 1$ treats in each party favor bag. If a total of $60x + 15$ treats are distributed, how many bags are given out?
   a. 15
   b. 16
   c. 20
   d. 22

5. Apples cost $2 each, while bananas cost $3 each. Maria purchased 10 fruits in total and spent $22. How many apples did she buy?
   a. 5
   b. 6
   c. 7
   d. 8

6. What are the polynomial roots of $x^2 + x - 2$?
   a. 1 and -2
   b. -1 and 2
   c. 2 and -2
   d. 9 and 13

7. What is the $y$-intercept of $y = x^{\frac{5}{3}} + (x - 3)(x + 1)$?
   a. 3.5
   b. 7.6
   c. -3
   d. -15.1

8. $x^4 - 16$ can be simplified to which of the following?

    a. $(x^2 - 4)(x^2 + 4)$
    b. $(x^2 + 4)(x^2 + 4)$
    c. $(x^2 - 4)(x^2 - 4)$
    d. $(x^2 - 2)(x^2 + 4)$

9. $(4x^2y^4)^{\frac{3}{2}}$ can be simplified to which of the following?

    a. $8x^3y^6$

    b. $4x^{\frac{5}{2}}y$

    c. $4xy$

    d. $32x^{\frac{7}{2}}y^{\frac{11}{2}}$

10. If $\sqrt{1 + x} = 4$, what is $x$?

    a. 10
    b. 15
    c. 20
    d. 25

11. Suppose $\frac{x+2}{x} = 2$. What is $x$?

    a. -1
    b. 0
    c. 2
    d. 4

12. A ball is thrown from the top of a high hill, so that the height of the ball as a function of time is $h(t) = -16t^2 + 4t + 6$, in feet. What is the maximum height of the ball in feet?

    a. 6
    b. 6.25
    c. 6.5
    d. 6.75

13. A rectangle has a length that is 5 feet longer than three times its width. If the perimeter is 90 feet, what is the length in feet?

    a. 10
    b. 20
    c. 25
    d. 35

14. Five students take a test. The scores of the first four students are 80, 85, 75, and 60. If the median score is 80, which of the following could NOT be the score of the fifth student?

    a. 60
    b. 80
    c. 85
    d. 100

15. In an office, there are 50 workers. A total of 60% of the workers are women, and the chances of a woman wearing a skirt is 50%. If no men wear skirts, how many workers are wearing skirts?

    a. 12

    b. 15

    c. 16

    d. 20

16. Ten students take a test. Five students get a 50. Four students get a 70. If the average score is 55, what was the last student's score?

    a. 20

    b. 40

    c. 50

    d. 60

17. A company invests $50,000 in a building where they can produce saws. If the cost of producing one saw is $40, then which function expresses the amount of money the company pays? The variable $y$ is the money paid and $x$ is the number of saws produced.

    a. $y = 50,000x + 40$

    b. $y + 40 = x - 50,000$

    c. $y = 40x - 50,000$

    d. $y = 40x + 50,000$

18. A six-sided die is rolled. What is the probability that the roll is 1 or 2?

    a. $\frac{1}{6}$

    b. $\frac{1}{4}$

    c. $\frac{1}{3}$

    d. $\frac{1}{2}$

19. A line passes through the origin and through the point (-3, 4). What is the slope of the line?

    a. $-\frac{4}{3}$

    b. $-\frac{3}{4}$

    c. $\frac{4}{3}$

    d. $\frac{3}{4}$

20. A pair of dice is thrown, and the sum of the two scores is calculated. What's the expected value of the roll?

    a. 5

    b. 6

    c. 7

    d. 8

21.

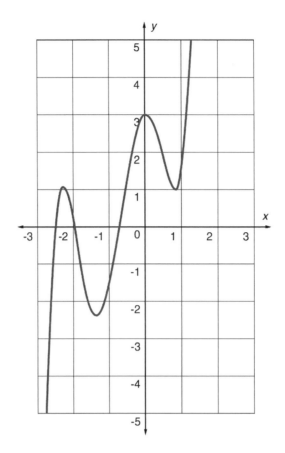

Which of the following functions represents the graph above?
a. $y = x^5 + 3.5x^4 - 6.5x^2 + 0.5x + 3$
b. $y = x^5 - 3.5x^4 + 6.5x^2 - 0.5x - 3$
c. $y = 5x^4 - 6.5x^2 + 0.5x + 3$
d. $y = -5x^4 - 6.5x^2 + 0.5x + 3$

22. Katie works at a clothing company and sold 192 shirts over the weekend. One third of the shirts that were sold were patterned, and the rest were solid. Which mathematical expression would calculate the number of solid shirts Katie sold over the weekend?
a. $192 \times \frac{1}{3}$

b. $192 \div \frac{1}{3}$

c. $192 \times (1 - \frac{1}{3})$

d. $192 \div 3$

23. Which measure for the center of a small sample set is most affected by outliers?
a. Mean
b. Median
c. Mode
d. None of the above

24. Given the value of a stock at monthly intervals, which graph should be used to best represent the trend of the stock?
    a. Box plot
    b. Line plot
    c. Line graph
    d. Circle graph

25. What is the probability of randomly picking the winner and runner-up from a race of 4 horses and distinguishing which is the winner?
    a. $\frac{1}{4}$

    b. $\frac{1}{2}$

    c. $\frac{1}{16}$

    d. $\frac{1}{12}$

26. What is the next number in the following series: $1, 3, 6, 10, 15, 21, \dots$ ?
    a. 26
    b. 27
    c. 28
    d. 29

27. A shipping box has a length of 8 inches, a width of 14 inches, and a height of 4 inches. If all three dimensions are doubled, what is the relationship between the volume of the new box and the volume of the original box?
    a. The volume of the new box is double the volume of the original box.
    b. The volume of the new box is four times as large as the volume of the original box.
    c. The volume of the new box is six times as large as the volume of the original box.
    d. The volume of the new box is eight times as large as the volume of the original box.

28. What is the product of the following expression?

$$(3 + 2i)(5 - 4i)$$

    a. $23 - 2i$
    b. $15 - 8i$
    c. $15 - 8i^2$
    d. $15 - 10i$

29.

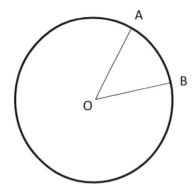

The length of arc $AB = 3\pi$ cm. The length of $\overline{OA} = 12$ cm. What is the degree measure of $\angle AOB$?
   a. 30 degrees
   b. 40 degrees
   c. 45 degrees
   d. 55 degrees

30.

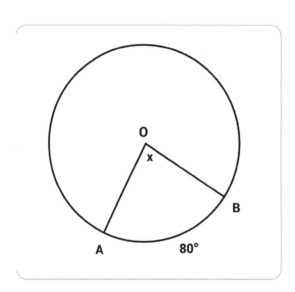

The area of circle O is $49\pi$ m. What is the area of the sector formed by $\angle AOB$?
   a. $80\pi$ m
   b. $10.9\pi$ m
   c. $4.9\pi$ m
   d. $10\pi$ m

31. The triangle shown below is a right triangle. What's the value of $x$?

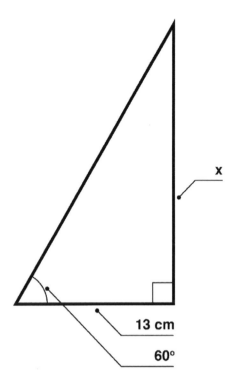

13 cm

60°

    a. $x = 1.73$
    b. $x = 0.57$
    c. $x = 13$
    d. $x = 22.52$

32. A ball is drawn at random from a ball pit containing 8 red balls, 7 yellow balls, 6 green balls, and 5 purple balls. What's the probability that the ball drawn is yellow?

    a. $\frac{1}{26}$

    b. $\frac{19}{26}$

    c. $\frac{7}{26}$

    d. 1

33. What's the probability of rolling a 6 at least once in two rolls of a die?

    a. $\frac{1}{3}$

    b. $\frac{1}{36}$

    c. $\frac{1}{6}$

    d. $\frac{11}{36}$

34. For a group of 20 men, the median weight is 180 pounds and the range is 30 pounds. If each man gains 10 pounds, which of the following would be true?
    a. The median weight will increase, and the range will remain the same.
    b. The median weight and range will both remain the same.
    c. The median weight will stay the same, and the range will increase.
    d. The median weight and range will both increase.

35. If the ordered pair $(-3, -4)$ is reflected over the $x$-axis, what's the new ordered pair?
    a. $(-3, -4)$
    b. $(3, -4)$
    c. $(3, 4)$
    d. $(-3, 4)$

36. If the volume of a sphere is $288\pi$ cubic meters, what are the radius and surface area of the same sphere?
    a. Radius 6 meters and surface area $144\pi$ square meters
    b. Radius 36 meters and surface area $144\pi$ square meter
    c. Radius 6 meters and surface area $12\pi$ square meters
    d. Radius 36 meters and surface area $12\pi$ square meters

37. Which four-sided shape is always a rectangle?
    a. Rhombus
    b. Square
    c. Parallelogram
    d. Quadrilateral

38. Using trigonometric ratios for a right angle, what is the value of the angle whose opposite side is equal to 25 centimeters and whose hypotenuse is equal to 50 centimeters?
    a. 15°
    b. 30°
    c. 45°
    d. 90°

39. Using trigonometric ratios for a right angle, what is the value of the closest angle whose adjacent side is equal to 7.071 centimeters and whose hypotenuse is equal to 10 centimeters?
    a. 15°
    b. 30°
    c. 45°
    d. 90°

40. Using trigonometric ratios, what is the value of an angle whose opposite side is equal to 1 inch and whose adjacent side is equal to the square root of 3 inches?
    a. 15°
    b. 30°
    c. 45°
    d. 90°

41. What is the function that forms an equivalent graph to $y = \cos(x)$?

    a. $y = \tan(x)$

    b. $y = \csc(x)$

    c. $y = \sin\left(x + \frac{\pi}{2}\right)$

    d. $y = \sin\left(x - \frac{\pi}{2}\right)$

42. A solution needs 5 mL of saline for every 8 mL of medicine given. How much saline is needed for 45 mL of medicine?

    a. $\frac{225}{8}$ mL

    b. 72 mL

    c. 28 mL

    d. $\frac{45}{8}$ mL

43. What's the midpoint of a line segment with endpoints $(-1, 2)$ and $(3, -6)$?

    a. $(1, 2)$

    b. $(1, 0)$

    c. $(-1, 2)$

    d. $(1, -2)$

44. A sample data set contains the following values: 1, 3, 5, 7. What's the standard deviation of the set?

    a. 2.58

    b. 4

    c. 6.23

    d. 1.1

## No Calculator Questions

45. An equilateral triangle has a perimeter of 18 feet. If a square whose sides have the same length as one side of the triangle is built, what will be the area of the square?

    a. 6 square feet

    b. 36 square feet

    c. 256 square feet

    d. 1000 square feet

46. What is the volume of a sphere, in terms of $\pi$, with a radius of 3 inches?

    a. $36\ \pi$ in$^3$

    b. $27\ \pi$ in$^3$

    c. $9\ \pi$ in$^3$

    d. $72\ \pi$ in$^3$

47. What is the length of the other leg of a right triangle with a hypotenuse of 10 inches and a leg of 8 inches?

    a. 6 in

    b. 18 in

    c. 80 in

    d. 13 in

48. A pizzeria owner regularly creates jumbo pizzas, each with a radius of 9 inches. She is mathematically inclined, and wants to know the area of the pizza to purchase the correct boxes and know how much she is feeding her customers. What is the area of the circle, in terms of $\pi$, with a radius of 9 inches?
   a. $81\,\pi$ in$^2$
   b. $18\,\pi$ in$^2$
   c. $90\,\pi$ in$^2$
   d. $9\,\pi$ in$^2$

49. How will the following expression be written in standard form?

$$(1 \times 10^4) + (3 \times 10^3) + (7 \times 10^1) + (8 \times 10^0)$$

   a. 137
   b. 13,078
   c. 1,378
   d. 8,731

50. What is the simplified form of the expression $tan\theta\ cos\theta$?
   a. $sin\theta$
   b. 1
   c. $csc\theta$
   d. $\dfrac{1}{sec\theta}$

51. What is the value of the sum of $\dfrac{1}{3}$ and $\dfrac{2}{5}$?
   a. $\dfrac{3}{8}$

   b. $\dfrac{11}{15}$

   c. $\dfrac{11}{30}$

   d. $\dfrac{4}{5}$

52. If the cosine of $30° = x$, the sine of what angle also equals $x$?
   a. 30°
   b. 60°
   c. 90°
   d. 120°

53. If the tangent of $45° = x$, the sine of what angle also equals $x$?
   a. 30°
   b. 60°
   c. 90°
   d. 120°

54. If $\overline{AE} = 4$, $\overline{AB} = 5$, and $\overline{AD} = 5$, what is the length of $\overline{AC}$?

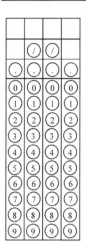

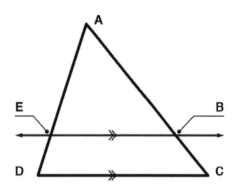

55. What is the decimal value of $\frac{3}{25}$?

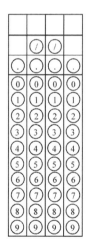

56. 6 is 30% of what number?

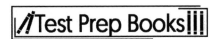
57. What is the value of the following expression?

$$\sqrt{8^2 + 6^2}$$

58. What is the measurement of angle f in the following picture? Assume the lines are parallel.

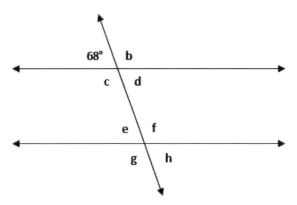

# Answer Explanations #3

**1. B:** First, subtract 4 from each side. This yields $6t = 12$. Now, divide both sides by 6 to obtain $t = 2$.

**2. B:** To be directly proportional means that $y = mx$. If $x$ is changed from 5 to 20, the value of $x$ is multiplied by 4. Applying the same rule to the y-value, also multiply the value of $y$ by 4. Therefore, $y = 12$.

**3. B:** From the slope-intercept form, $y = mx + b$, it is known that $b$ is the $y$-intercept, which is 1. Compute the slope as $\frac{2-1}{1-0} = 1$, so the equation should be $y = x + 1$.

**4. A:** Each bag contributes $4x + 1$ treats. The total treats will be in the form $4nx + n$ where $n$ is the total number of bags. The total is in the form $60x + 15$, from which it is known that $n = 15$.

**5. D:** Let $a$ be the number of apples and $b$ the number of bananas. Then, the total cost is $2a + 3b = 22$, while it also known that $a + b = 10$. Using the knowledge of systems of equations, cancel the $b$ variables by multiplying the second equation by -3. This makes the equation $-3a - 3b = -30$. Adding this to the first equation, the $b$ values cancel to get $-a = -8$, which simplifies to $a = 8$.

**6. A:** Finding the roots means finding the values of $x$ when $y$ is zero. The quadratic formula could be used, but in this case it is possible to factor by hand, since the numbers -1 and 2 add to 1 and multiply to -2. So, factor:

$$x^2 + x - 2 = (x - 1)(x + 2) = 0$$

then set each factor equal to zero. Solving for each value gives the values $x = 1$ and $x = -2$.

**7. C:** To find the $y$-intercept, substitute zero for $x$, which gives us:

$$y = 0^{\frac{5}{3}} + (0 - 3)(0 + 1)$$

$$0 + (-3)(1)$$

$$-3$$

**8. A:** This has the form $t^2 - y^2$, with $t = x^2$ and $y = 4$. It's also known that $t^2 - y^2 = (t + y)(t - y)$, and substituting the values for $t$ and $y$ into the right-hand side gives:

$$(x^2 - 4)(x^2 + 4)$$

**9. A:** Simplify this to:

$$(4x^2y^4)^{\frac{3}{2}} = 4^{\frac{3}{2}}(x^2)^{\frac{3}{2}}(y^4)^{\frac{3}{2}}$$

Now, simplify the numeric term in the expression:

$$4^{\frac{3}{2}} = (\sqrt{4})^3 = 2^3 = 8$$

For the other, recall that the exponents must be multiplied, so this yields:

$$8x^{2\times\frac{3}{2}}y^{4\times\frac{3}{2}} = 8x^3y^6$$

**10. B:** Start by squaring both sides to get $1 + x = 16$. Then subtract 1 from both sides to get $x = 15$.

**11. C:** Multiply both sides by $x$ to get $x + 2 = 2x$, which simplifies to $-x = -2$, or $x = 2$.

**12. B:** The independent variable's coordinate at the vertex of a parabola (which is the highest point, when the coefficient of the squared independent variable is negative) is given by $x = -\dfrac{b}{2a}$.

Substitute and solve for $x$ to get:

$$x = -\frac{4}{2(-16)} = \frac{1}{8}$$

Using this value of $x$, the maximum height of the ball ($y$), can be calculated. Substituting $x$ into the equation yields:

$$h(t) = -16\left(\frac{1}{8}\right)^2 + 4\left(\frac{1}{8}\right) + 6 = 6.25$$

**13. D:** Denote the width as $w$ and the length as $l$. Then, $l = 3w + 5$. The perimeter is $2w + 2l = 90$. Substituting the first expression for $l$ into the second equation yields:

$$2(3w + 5) + 2w = 90$$

$$6w + 10 + 2w = 90$$

$$8w = 80$$

$$w = 10$$

Putting this into the first equation, it yields:

$$l = 3(10) + 5 = 35$$

**14. A:** Lining up the given scores provides the following list: 60, 75, 80, 85, and one unknown. Because the median needs to be 80, it means 80 must be the middle data point out of these five. Therefore, the unknown data point must be the fourth or fifth data point, meaning it must be greater than or equal to 80. The only answer that fails to meet this condition is 60.

**15. B:** If 60% of 50 workers are women, then there are 30 women working in the office. If half of them are wearing skirts, then that means 15 women wear skirts. Since none of the men wear skirts, this means there are 15 people wearing skirts.

**16. A:** Let the unknown score be $x$. The average will be:

$$\frac{5 \times 50 + 4 \times 70 + x}{10} = \frac{530 + x}{10} = 55$$

Multiply both sides by 10 to get $530 + x = 550$, or $x = 20$.

**17. D:** For manufacturing costs, there is a linear relationship between the cost to the company and the number produced, with a $y$-intercept given by the base cost of acquiring the means of production, and a slope given by the cost to produce one unit. In this case, that base cost is $50,000, while the cost per unit is $40. So:

$$y = 40x + 50,000$$

**18. C:** A die has an equal chance for each outcome. Since it has six sides, each outcome has a probability of $\frac{1}{6}$. The chance of a 1 or a 2 is therefore:

$$\frac{1}{6} + \frac{1}{6} = \frac{1}{3}$$

**19. A:** The slope is given by:

$$m = \frac{y_2 - y_1}{x_2 - x_1}$$

$$\frac{0 - 4}{0 - (-3)}$$

$$-\frac{4}{3}$$

**20. C:** The expected value is equal to the total sum of each product of individual score and probability. There are 36 possible rolls. The probability of rolling a 2 is $\frac{1}{36}$. The probability of rolling a 3 is $\frac{2}{36}$. The probability of rolling a 4 is $\frac{3}{36}$. The probability of rolling a 5 is $\frac{4}{36}$. The probability of rolling a 6 is $\frac{5}{36}$. The probability of rolling a 7 is $\frac{6}{36}$. The probability of rolling an 8 is $\frac{5}{36}$. The probability of rolling a 9 is $\frac{4}{36}$. The probability of rolling a 10 is $\frac{3}{36}$. The probability of rolling an 11 is $\frac{2}{36}$. Finally, the probability of rolling a 12 is $\frac{1}{36}$.

Each possible outcome is multiplied by the probability of it occurring. Like this:

$$2 \times \frac{1}{36} = a$$

$$3 \times \frac{2}{36} = b$$

$$4 \times \frac{3}{36} = c$$

And so forth.

Then all of those results are added together:

$$a + b + c \ldots = expected\ value$$

In this case, it equals 7.

**21. A:** The graph contains four turning points (where the curve changes from rising to falling or vice versa). This indicates that the degree of the function (highest exponent for the variable) is 5, eliminating Choices C and D. The $y$-intercepts of the functions can be determined by substituting 0 for $x$ and finding the value of $y$. The function for Choice A has a $y$-intercept of 3, and the function for Choice B has a $y$-intercept of -3. Therefore, Choice B is eliminated.

**22. C:** $\frac{1}{3}$ of the shirts sold were patterned. Therefore, $1 - \frac{1}{3} = \frac{2}{3}$ of the shirts sold were solid. Anytime "of" a quantity appears in a word problem, multiplication should be used. Therefore:

$$192 \times \frac{2}{3}$$

$$\frac{192 \times 2}{3}$$

$$\frac{384}{3}$$

128 solid shirts were sold

The entire expression is:

$$192 \times \left(1 - \frac{1}{3}\right)$$

**23. A:** Mean. An outlier is a data value that is either far above or far below the majority of values in a sample set. The mean is the average of all the values in the set. In a small sample set, a very high or very low number could drastically change the average of the data points. Outliers will have no more of an effect on the median (the middle value when arranged from lowest to highest) than any other value above or below the median. If the same outlier does not repeat, outliers will have no effect on the mode (value that repeats most often).

**24. C:** Line graph. The scenario involves data consisting of two variables, month and stock value. Box plots display data consisting of values for one variable. Therefore, a box plot is not an appropriate choice. Both line plots and circle graphs are used to display frequencies within categorical data. Neither can be used for the given scenario. Line graphs display two numerical variables on a coordinate grid and show trends among the variables.

**25. D:** $\frac{1}{12}$. The probability of picking the winner of the race is $\frac{1}{4}$ or:

$$\left(\frac{number\ of\ favorable\ outcomes}{number\ of\ total\ outcomes}\right)$$

Assuming the winner was picked on the first selection, three horses remain from which to choose the runner-up (these are dependent events). Therefore, the probability of picking the runner-up is $\frac{1}{3}$. To determine the probability of multiple events, the probability of each event is multiplied:

$$\frac{1}{4} \times \frac{1}{3} = \frac{1}{12}$$

**26. C:** Each number in the sequence is adding one more than the difference between the previous two.

For example, $10 - 6 = 4, 4 + 1 = 5$.

Therefore, the next number after 10 is $10 + 5 = 15$.

Going forward, $21 - 15 = 6, 6 + 1 = 7$. The next number is $21 + 7 = 28$. Therefore, the difference between numbers is the set of whole numbers starting at 2: 2, 3, 4, 5, 6, 7....

**27. D:** The formula for finding the volume of a rectangular prism is $V = l \times w \times h$ where $l$ is the length, $w$ is the width, and $h$ is the height. The volume of the original box is calculated:

$$V = 8 \times 14 \times 4 = 448 \text{ in}^3$$

The volume of the new box is calculated:

$$V = 16 \times 28 \times 8 = 3584 \text{ in}^3$$

The volume of the new box divided by the volume of the old box equals 8.

**28. A:** The notation $i$ stands for an imaginary number. The value of $i$ is equal to $\sqrt{-1}$. When performing calculations with imaginary numbers, treat $i$ as a variable, and simplify when possible. Multiplying the binomials by the FOIL method produces:

$$15 - 12i + 10i - 8i^2$$

Combining like terms yields:

$$15 - 2i - 8i^2$$

Since $i = \sqrt{-1}$, $i^2 = (\sqrt{-1})^2 = -1$.

Therefore, substitute -1 for $i^2$:

$$15 - 2i - 8(-1)$$

Simplifying results in:

$$15 - 2i + 8 = 23 - 2i$$

**29. C:** The formula to find arc length is $s = \theta r$ where $s$ is the arc length, $\theta$ is the radian measure of the central angle, and $r$ is the radius of the circle. Substituting the given information produces: $3\pi$ cm $= \theta 12$ cm. Solving for $\theta$ yields $\theta = \frac{\pi}{4}$. To convert from radian to degrees, multiply the radian measure by $\frac{180°}{\pi}$:

$$\frac{\pi}{4} \times \frac{180°}{\pi} = 45°$$

**30. B:** Given the area of the circle, the radius can be found using the formula $A = \pi r^2$. In this case, $49\pi = \pi r^2$, which yields $r = 7$ m. A central angle is equal to the degree measure of the arc it inscribes; therefore, $\angle x = 80°$. The area of a sector can be found using the formula:

$$A = \frac{\theta}{360°} \times \pi r^2$$

In this case:

$$A = \frac{80^{\circ}}{360^{\circ}} \times \pi(7)^2 = 10.9\pi \text{ m}$$

**31. D:** SOHCAHTOA is used to find the missing side length. Because the angle and adjacent side are known, $\tan 60 = \frac{x}{13}$.

Making sure to evaluate tangent with an argument in degrees, this equation gives"

$$x = 13 \tan 60 = 13 \times \sqrt{3} = 22.52$$

**32. C:** The sample space is made up of $8 + 7 + 6 + 5 = 26$ balls.

The probability of pulling each individual ball is $\frac{1}{26}$. Since there are 7 yellow balls, the probability of pulling a yellow ball is $\frac{7}{26}$.

**33. D:** The addition rule is necessary to determine the probability because a 6 can be rolled on either roll of the die. The rule used is:

$$P(A \text{ or } B) = P(A) + P(B) - P(A \text{ and } B)$$

The probability of a 6 being individually rolled is $\frac{1}{6}$ and the probability of a 6 being rolled twice is:

$$\frac{1}{6} \times \frac{1}{6} = \frac{1}{36}$$

Therefore, the probability that a 6 is rolled at least once is:

$$\frac{1}{6} + \frac{1}{6} - \frac{1}{36} = \frac{11}{36}$$

**34. A:** If each man gains 10 pounds, every original data point will increase by 10 pounds. Therefore, the man with the original median will still have the median value, but that value will increase by 10. The smallest value and largest value will also increase by 10 and, therefore, the difference between the two won't change. The range does not change in value and, thus, remains the same.

**35. D:** When an ordered pair is reflected over an axis, the sign of one of the coordinates must change. When it's reflected over the $x$-axis, the sign of the $y$-coordinate must change. The $x$-value remains the same. Therefore, the new ordered pair is $(-3, 4)$.

**36. A:** Because the volume of the given sphere is $288\pi$ cubic meters, this gives:

$$\frac{4}{3}\pi r^3 = 288\pi$$

This equation is solved for $r$ to obtain a radius of 6 meters. The formula for surface area is $4\pi r^2$ so:

$$SA = 4\pi6^2 = 144\pi \text{ square meters}$$